Disclaimer

The publisher of this book is by no way associated with the National Institute of Standards and Technology (NIST). The NIST did not publish this book. It was published by 50 page publications under the public domain license.

50 Page Publications.

Book Title: Enhancement of the Virtual Cybernetic Building Testbed to Include a Zone Fire Model With HVAC Components

Book Author: Cheol D. Park; Paul A. Reneke; Michael A. Galler; Steven T. Bushby; William D. Davis

Book Abstract: As part of a Cybernetic Building Systems project at the NIST Building and Fire Research Laboratory, a whole building emulator called the Virtual Cybernetic Building Testbed (VCBT) has been developed. The VCBT combines simulation tools with commercial building automation system products in order to emulate the characteristics of a commercial building and its various systems and controls. The VCBT simulates in real time the building envelope, weather, the heating ventilating and air-conditioning (HVAC) system, the heating/cooling plant, the fire detection system, and the security system. The simulation is linked through a data acquisition system to commercial products for HVAC control, fire detection, access control and lighting control in order to study their performance and interactions. This report describes a new simulation tool developed for the VCBT that combines a fire simulator called the Zone Fire Model with building shell and HVAC components derived from HVACSIM+. This new simulation tool is called the Zone Fire Model with HVAC components (ZFM-HVAC). Implementation of the ZFM-HVAC tool in the VCBT is described along with the results of emulation runs used to test its **performance.**

Citation: NIST Interagency/Internal Report (NISTIR) - 7414

Keyword: BACnet;building automation and control;building management systems;cybernetic building systems;fire models;fire simulator;ZEM-HVAC

NISTIR 7414

Enhancement of the Virtual Cybernetic Building Testbed to Include a Zone Fire Model with HVAC Components

Cheol Park
Paul A. Reneke
Michael A. Galler
Steven T. Bushby
William D. Davis

National Institute of Standards and Technology
Technology Administration, U.S. Department of Commerce

NISTIR 7414

Enhancement of the Virtual Cybernetic Building Testbed to Include a Zone Fire Model with HVAC Components

Cheol Park
Paul A. Reneke
Michael A. Galler
Steven T. Bushby
William D. Davis

Building and Fire Research Laboratory
Gaithersburg, MD 20899

April 2007

U.S. Department of Commerce
Carlos M. Gutierrez, Secretary

Technology Administration
Robert Cresanti, Under Secretary of Commerce for Technology

National Institute of Standards and Technology
William Jeffrey, Director

ABSTRACT

Enhancement of the Virtual Cybernetic Building Testbed to Include a Zone Fire Model with HVAC Components

Cheol Park, Paul A. Reneke, Michael A. Galler, Steven T. Bushby, and William D. Davis
Building and Fire Research Laboratory, NIST, Gaithersburg, MD 20899

As part of a Cybernetic Building Systems project at the NIST Building and Fire Research Laboratory, a whole building emulator called the Virtual Cybernetic Building Testbed (VCBT) has been developed. The VCBT combines simulation tools with commercial building automation system products in order to emulate the characteristics of a commercial building and its various systems and controls. The VCBT simulates in real time the building envelope, weather, the heating ventilating and air-conditioning (HVAC) system, the heating/cooling plant, the fire detection system, and the security system. The simulation is linked through a data acquisition system to commercial products for HVAC control, fire detection, access control and lighting control in order to study their performance and interactions.

This report describes a new simulation tool developed for the VCBT that combines a fire simulator called the Zone Fire Model with building shell and HVAC components derived from $HVACSIM^{+}$. This new simulation tool is called the Zone Fire Model with HVAC components (ZFM-HVAC). Implementation of the ZFM-HVAC tool in the VCBT is described along with the results of emulation runs used to test its performance.

Key words: BACnet, building automation and control, building management systems, cybernetic building systems, fire models, fire simulator, ZFM-HVAC

Table of Contents

1. INTRODUCTION 1
2. VCBT 2.0 ARCHITECTURE 2
   - 2.1 Center 3
   - 2.2 HVACSIM$^+$ 5
   - 2.3 Building Shell Model 7
   - 2.4 Data Acquisition System 8
   - 2.5 Building System BACnet Controllers 10
   - 2.6 BACnet Fire Panel 13
   - 2.7 Biometric Devices for Security 14
   - 2.8 Sensor-Driven Fire Model (SDFM) 14
3. ZFM-HVAC 14
   - 3.1 Zone Fire Model (ZFM) 15
   - 3.2 HVAC Components from HVACSIM$^+$ 15
     - 3.2.1 Actuator (Type321) 17
     - 3.2.2 Mixing box 19
     - 3.2.3 Temperature Sensor (Type301) 20
     - 3.2.4 Humidity Sensor (Type302) 21
     - 3.2.5 Flow rate Sensor (Type303) 22
     - 3.2.6 Static pressure Sensor (Type305) 22
     - 3.2.7 Motor drive (Type333) 23
     - 3.2.8 Fan (Type350) 23
     - 3.2.9 VAV Box (Type526) 24
     - 3.2.10 Fluid Resistance (Type341) 26
     - 3.2.11 Add Heat (Type366) 26
     - 3.2.12 Merge Flow (Type422) 26
     - 3.2.13 Mix Flow (Type365) 27
     - 3.2.14 Mass Flow Balance (Type425) 27
     - 3.2.15 Coil & Valve (Type524) 27
4. ZFM-HVAC RUNS AND DISCUSSION 35
   - 4.1 Case 1 Scenario 36
   - 4.2 Case 2 Scenario 57
   - 4.3 Case 3 Scenario 62
5. CONCLUSION 68
6. REFERENCES 69

Appendix A. Sample BACnet Air-handling Unit Controller Object Database 71

Appendix B. Sample BACnet VAV Box Controller Object Database........................72

Appendix C. Fire-related Input Data for Case 1_1..73

Appendix D. HVAC-side Configuration Input Data..74

Appendix E. Sample AHU Input Data...75

Appendix F. Sample VAV Input Data...83

Appendix G. Sample AHU Initial Data..86

1. INTRODUCTION

As part of a Cybernetic Building Systems project at the Building and Fire Research Laboratory of National Institute of Standards and Technology (NIST), a whole building emulator called the Virtual Cybernetic Building Testbed (VCBT) has been developed. The VCBT combines simulation tools with commercial building automation system products in order to emulate the characteristics of a commercial building and its various systems and controls. The VCBT simulates in real time the building envelope, weather, the heating ventilating and air-conditioning (HVAC) system, the heating/cooling plant, the fire detection system, and the security system. The simulation is linked through a data acquisition system to commercial products for HVAC control, fire detection, access control and lighting control in order to study their performance and interactions. Simulation results are converted by the data acquisition system into electrical signals that appear to building automation system devices as input from real sensors. The control actions are captured and used by the simulation software to determine the response of the virtual building. The overall effect is that the building management system (BMS) controls the simulated building systems as if they were an actual building.

In 1983, a simple building emulator was constructed at NIST to demonstrate the basic concept of a building emulator. The first emulator had a model that was too simple for testing algorithms quantitatively. Later a contractor for the U.S. Naval Civil Engineering Laboratory, with assistance from NIST, built a second building emulator with greater sophistication [1]. Since then the work on building emulators has continuously evolved [2-7]. The resulting VCBT 1.0 simulator employed the program, HVACSIM$^+$ [8], which stands for HVAC SIMulation PLUS other systems to simulate the HVAC system components and simple building envelopes except for the BMS controllers. The VCBT has been used to study various types of building systems and technology including: interactions of HVAC system components with fire, HVAC system fault detection and diagnostics (FDD) technology [9], communication protocols for building automation and control networks (BACnet) [10, 11], and the sensor-driven fire model (SDFM) [12].

Since the VCBT 1.0 architecture was reported [7], several enhancements have been made including: a more advanced fire detection system panel, a biometric building access control system, newer HVAC controls, and a new simulation tool developed for the VCBT that combines a fire simulator called the Zone Fire Model with building shell and HVAC components derived from HVACSIM$^+$. This new simulation tool is called the Zone Fire Model with HVAC components (ZFM-HVAC).

This report reviews the overall structure of the VCBT and describes the implementation of the ZFM-HVAC tool including results of emulation runs used to test its performance. Details of the ZFM components are published elsewhere [12]. Details of the HVAC components used in ZFM-HVAC will be described in this report.

2. VCBT 2.0 ARCHITECTURE

The architecture of the VCBT version 2.0 is shown schematically in Figure 1. The VCBT consists of a Center that provides an overall control and coordinating function of the other components, some of which are implemented in software and others are real building automation system devices. Simulated components include weather data, a simple building envelope model implemented using HVACSIM$^+$ called the Shell Model, a firefighter decision support tool called the Sensor-Driven Fire Model (SDFM), and a visualization tool that can depict the virtual building and its internal conditions implemented in Virtual Reality Modeling Language (VRML). Two options exist for simulating the building interior rooms and HVAC system. HVACSIM$^+$ is used if the research requires very detailed simulation of HVAC system components. ZFM-HVAC is used if the research requires investigating fires in the building. Fire detectors are represented by a combination of simulated output from detectors and actual fire detection hardware.

Physical components of the VCBT include a variety of BACnet HVAC controllers from several manufacturers, a BACnet fire alarm panel, and a biometric building access control system. The hardware components are linked to the software components through a commercial data acquisition system. Key simulated values are converted by the data acquisition system into voltage or current signals that are wired to the building automation system devices and appear as sensor inputs. Control actions are captured by the data acquisition system and converted to digital values that are fed into the next step of the simulation. Figure 2 is a photo of the VCBT hardware.

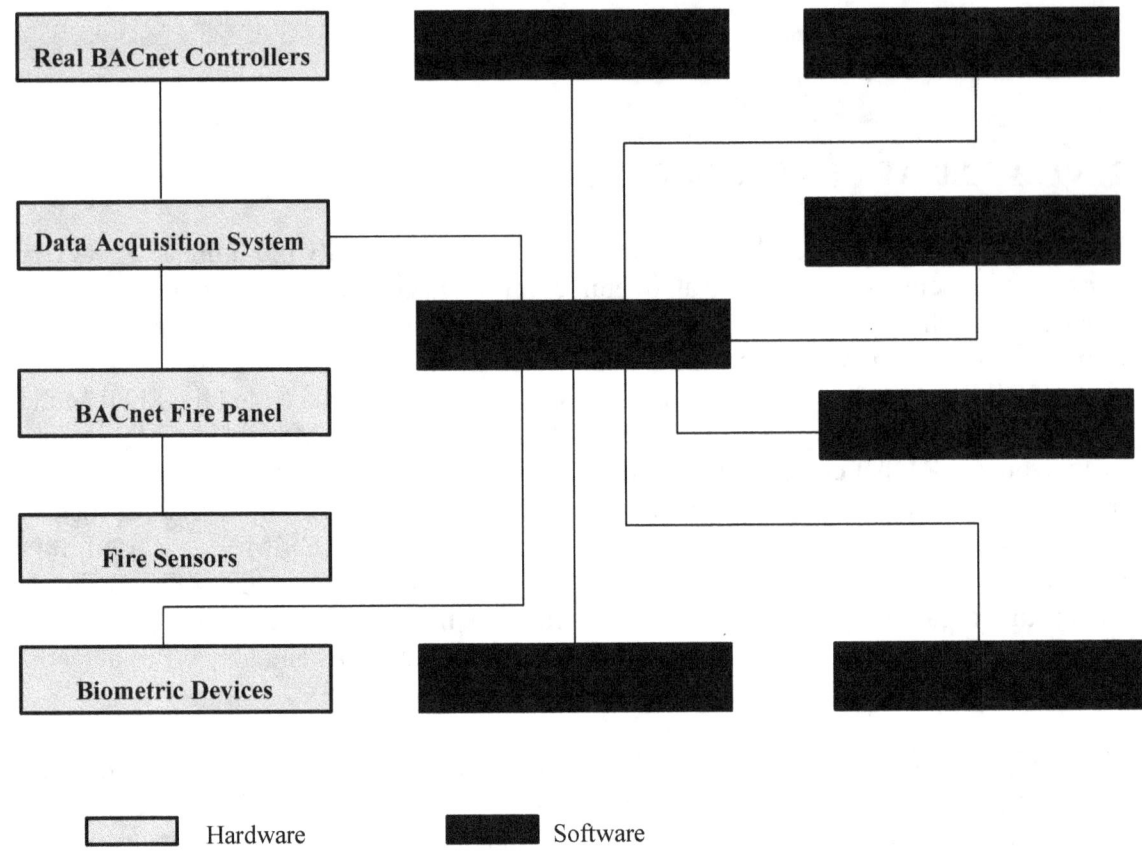

Figure 1. Components of the VCBT distributed simulation environment.

2.1 Center

The Center is the heart of the VCBT. It is a computer program residing in a personal computer that uses the Common Object Request Broker Architecture (CORBA) [13] to provide communications between the distributed components of the VCBT. CORBA is a vendor-independent, object-oriented architecture that provides a way to link computer programs across a network.

Using the graphical interface of the Center, configuration data are entered and output data from the simulations are displayed. The inputs and outputs from the building control system hardware are also monitored by the Center. The VCBT runs in real-time in a distributed, multi-platform environment. As seen in Figure 1, the Center communicates with the building management system hardware through the data acquisition system. Other components of the VCBT communicate directly with the Center using CORBA.

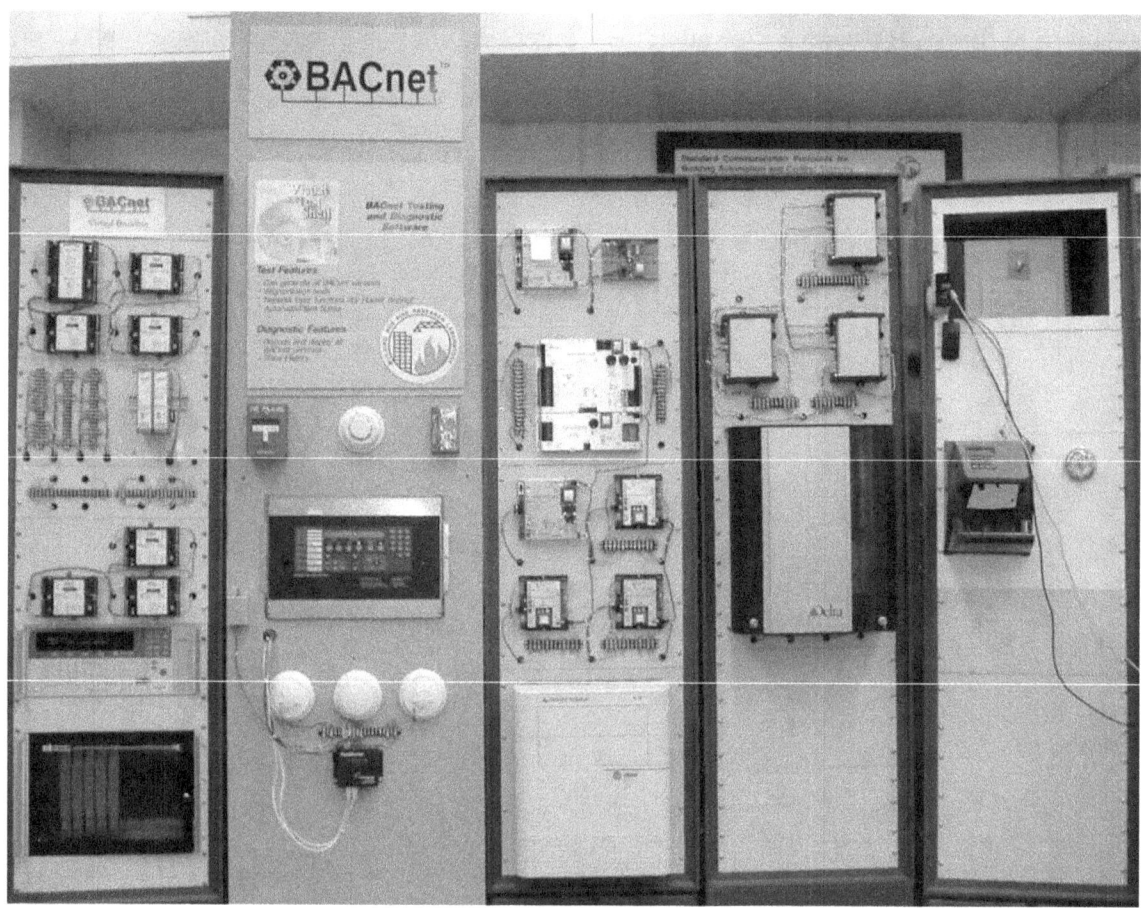

Figure 2. VCBT controller hardware, data acquisition system, fire-related devices including fire-panel, and biometric devices. Workstations for the VCBT controllers are not shown.

The Center controls the flow of data between all elements of the VCBT. It is also responsible for starting, initializing, and closing them. Individual components can be selected for use in a particular run of the VCBT, and the computer that these components will run on can be chosen at runtime. The Center runs on a 10 s cycle, during which data messages for each active component are created, transmitted, and any return messages evaluated. The amount of data currently being sent through the Center is only a fraction of its capability.

The Center is responsible for storing every piece of data that is transferred between the various components of the VCBT. This is done by first creating a list of the objects (i.e., a variable air volume (VAV) box or a room) which have properties (i.e., temperature, humidity, set points) associated with them. The properties of each object are then created, and linked to their parent objects. For each component of the VCBT, the subset of object and property combinations that the component uses is stored in a list associated with that component. Separate lists are created for each component for data going to the component, and for data coming back from the component. The value of any property used in the VCBT can be viewed at the Center. Figure 3 is a screen image of the Center.

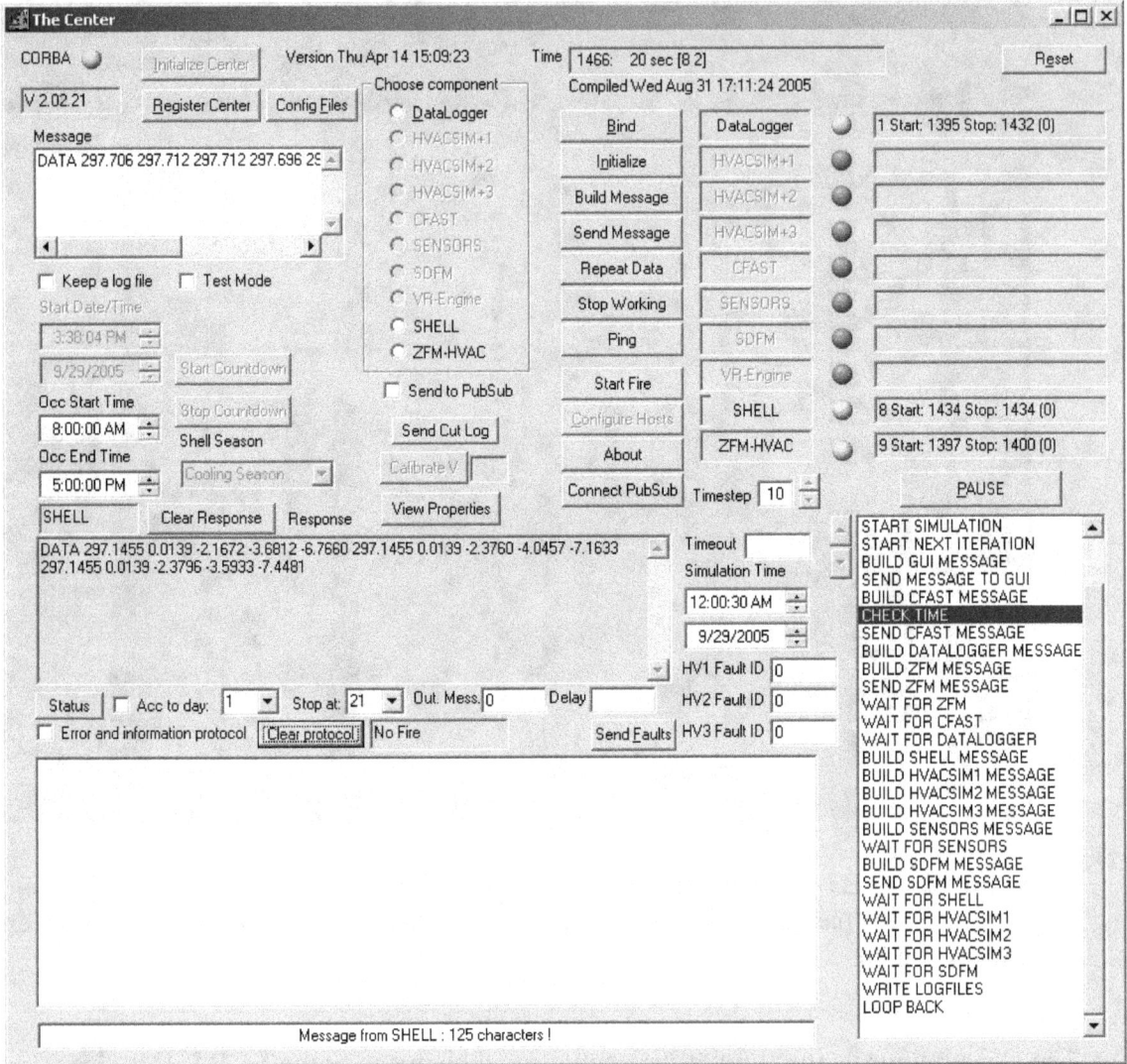

Figure 3. The Center Interface is used to link the components of a particular emulation and to control their interaction. In this example only the Datalogger, the ZFM-HVAC and the shell components are used.

2.2 HVACSIM$^+$

HVACSIM$^+$ is a public domain computer simulation program developed by NIST for studying the dynamic interactions between building system components [8]. Most of the programs are written in Fortran 77. This program has been used extensively in the studies of fault detection and diagnostics (FDD) of building HVAC systems [9].

HVACSIM⁺ links component type models together to build an HVAC system simulation. The component models used in the VCBT were originally developed during the course of ASHRAE Research Project 825 [14]. A typical virtual building simulation consists of the building zones (rooms), some single-duct Variable Air Volume (VAV) air-handling units, VAV terminal boxes, and the associated valves, coils, dampers, and sensors.

A schematic representation of a typical VCBT air-handling unit along with central plant facilities is shown in Figure 4. The figure illustrates the various dampers, valves, actuators, and fans that are to be controlled. The simulated sensor outputs that are to be used as inputs to the controllers for air-handling units are listed in Table 1. The component models of a central boiler, chiller, and cooling tower have not been utilized in the VCBT yet and are shown as dashed lines. Figure 5 is a schematic drawing of one of the single duct VAV boxes.

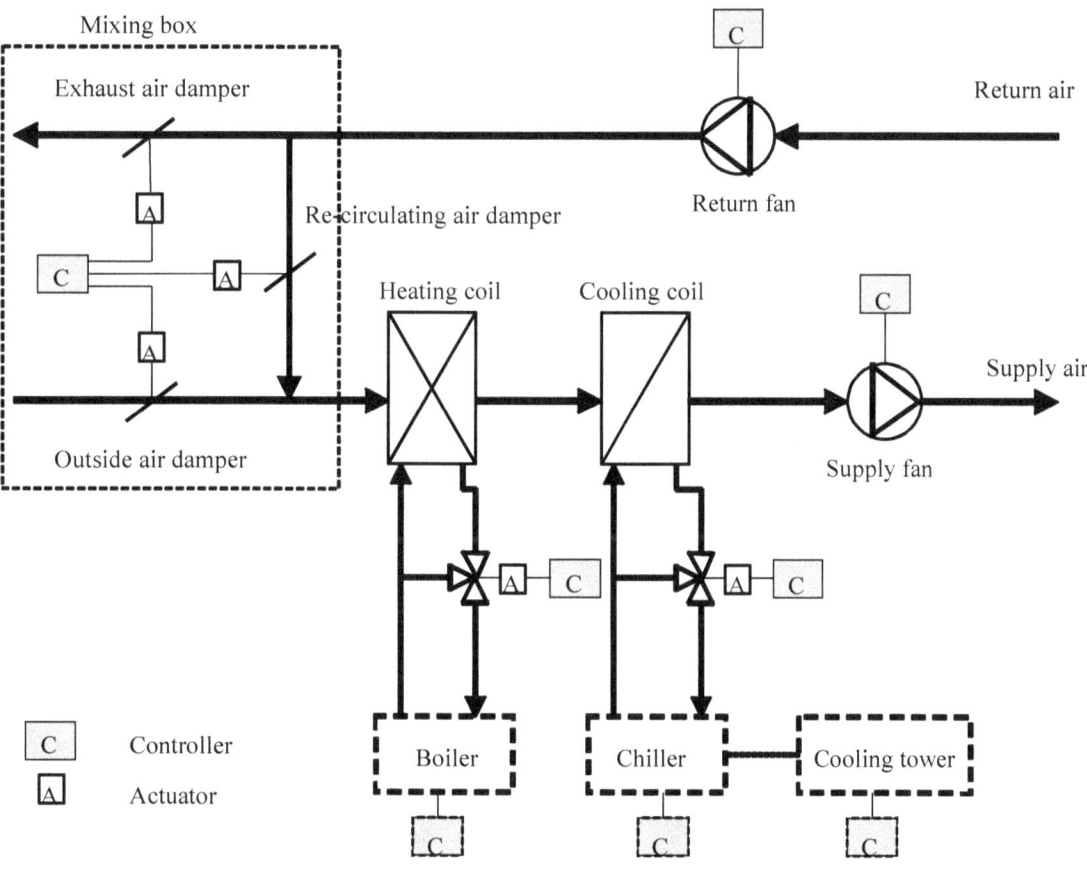

Figure 4. A schematic representation of a VCBT air-handling unit.

Table 1. Simulated sensor outputs for Air-handling unit control.

mixed air temperature
outdoor air relative humidity
outdoor air temperature
return air relative humidity
return air temperature
return air volumetric flow rate
supply air static pressure
supply air temperature
supply air volumetric flow rate

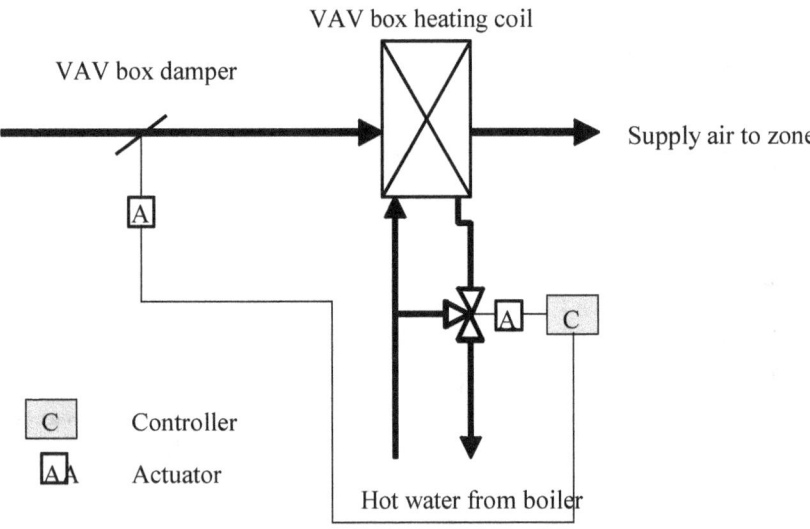

Figure 5. A schematic representation of the single duct VAV box.

2.3 Building Shell Model

The building shell model component of the VCBT is implemented using HVACSIM$^+$. The building shell model simulates heat gain or loss through the building envelope surfaces (walls, ceilings, floors, doors, and windows) as a function of outside conditions such as air temperatures, solar intensities, and wind speeds. Conduction heat transfer functions are used for calculation of heat fluxes through composite multi-layered surfaces in which each layer material is assumed to have uniform properties [15]. By varying the input weather data, the climate effects due to building location can be observed.

Table 2 lists the variables associated with a typical building surface. The VCBT shell model has been enhanced to use ASHRAE WYEC2 hourly weather data [16]. The shell model assumes that weather conditions remain constant for 15-minute intervals. The 15-minute data are generated from the hourly values by interpolation. February, July, and October are selected as representatives of three weather patterns for heating, cooling, and swing seasons, respectively. Internal load variations and solar heat gains through windows and facades are not yet considered in the current ZFM-HVAC, but the temperature and humidity of outside air are accounted for.

Table 2. Building Surface Variables.

Variable	Description
Inputs	
TIA	Indoor air dry-bulb temperature
TMR	Mean radiant temperature
TOSINF	Outer surface temperature of unexposed wall
FAHADW	Shaded fraction of exposed surface
Outputs	
TIS	Inner surface temperature
SOLINT	Integrated solar influx on surface
Parameters	
IZN	Identification number of zone
ID	Identification number of surface
IEXPOS	Internal exposure 0 = within zone, 1= between zones, 2=exposed to sun
ISTR	Identification number of the construct
AS	Surface area
ORIENT	Azimuth angle of normal to surface & south
TILT	Tilt angle: flat roof=0^E, floor=180^E
GRF	Ground reflectivity
IROFS	Outer surface roughness index; 1=stucco, ...
ABSOS	Solar absorptance of outer surface
ABSIS	Short wave absorptance of inner surface
EMITIS	Emissivity of the inner surface
TRANSM	Transmittance of the glass window
SC	Shading coefficient of the glass window

2.4 Data Acquisition System

One of the computers used for the distributed simulation contains an interface card that communicates digitally with a commercial data acquisition system. This component is called the Datalogger in the system architecture. The simulated values that represent sensor inputs to the controllers are converted by the digital-to-analog converter of the data acquisition system into analog outputs represented by either DC voltages (0 to 5 V) or DC currents (4 to 20 ma). There is a capability to scale values to other ranges if needed. The output values of the BACnet controllers that represent inputs to the virtual building system component models are read by the digital voltmeter in conjunction with multiplexing scanners. The output voltage range of each controller is in between 0 V and 10 V. Solid-state relay chips are employed for handling binary (digital) signals between a BACnet controller and the data acquisition system.

Figure 6 shows the relationship between the data acquisition system with other components of the VCBT. "Hardwire" represents a pair of electrical wires. Other connections are digital communication using CORBA. After receiving data from the simulations, converting it into the analog values, and sending it to the controllers, the Datalogger waits 1 s, and then begins to read the controller outputs. The total cycle time is approximately 3 s to 4 s.

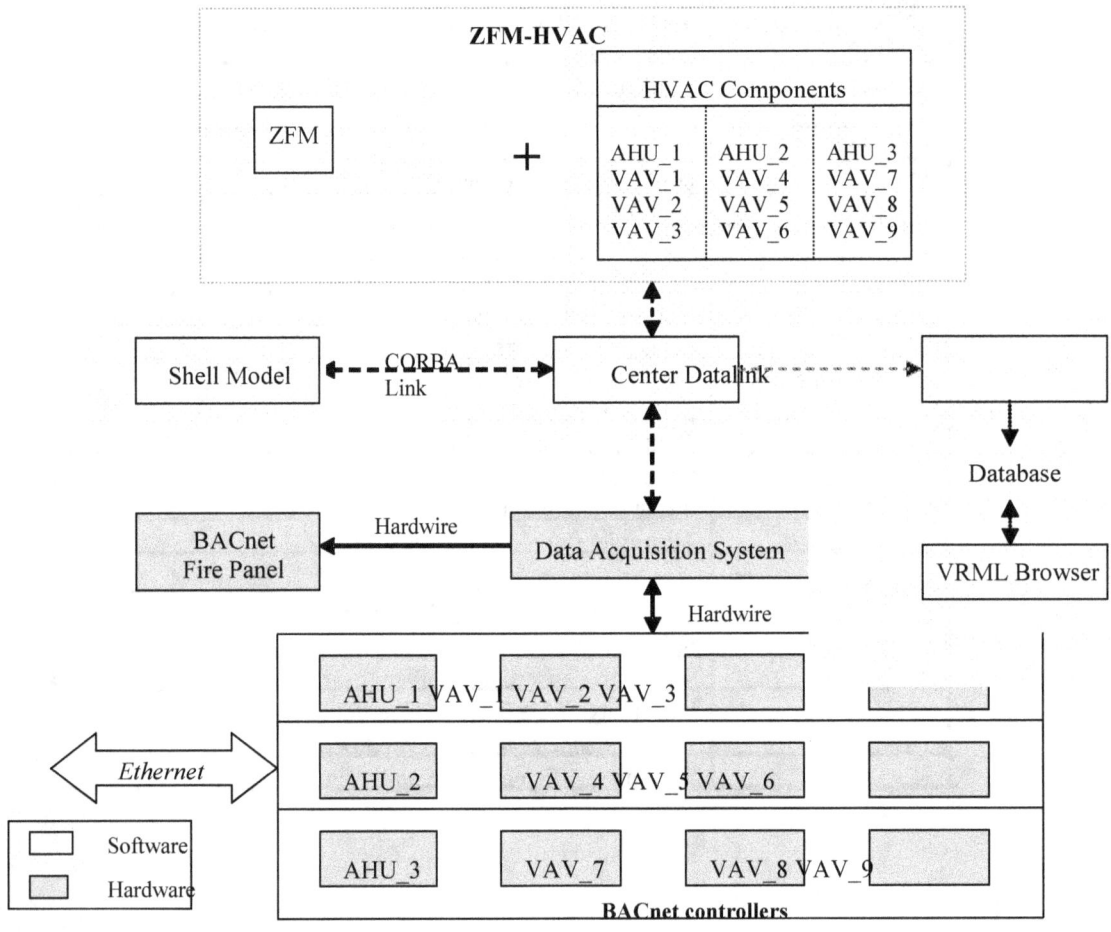

Figure 6. Relationship of the data acquisition system for ZFM-HVAC with other units.

2.5 Building System BACnet Controllers

Commercially available BACnet controllers make up the building automation and control system of the virtual building. The controller input and output connections are wired to a patch panels that provides connections to the emulator. Various BACnet communication networks [10, 11] connect the controllers to each other.

Figure 7 shows the network topology of the current BACnet system. For each local area network (LAN), the figure shows the type of network technology used and the BACnet network number. An Ethernet LAN serves as the backbone to the system. It connects three operator workstations and the supervisory controllers for the virtual building. These supervisory controllers provide scheduling capability, alarm processing, implementing reset schedules and also serve as routers to networks of unitary controllers. The unitary controllers reside on ARCNET or MS/TP LANs.

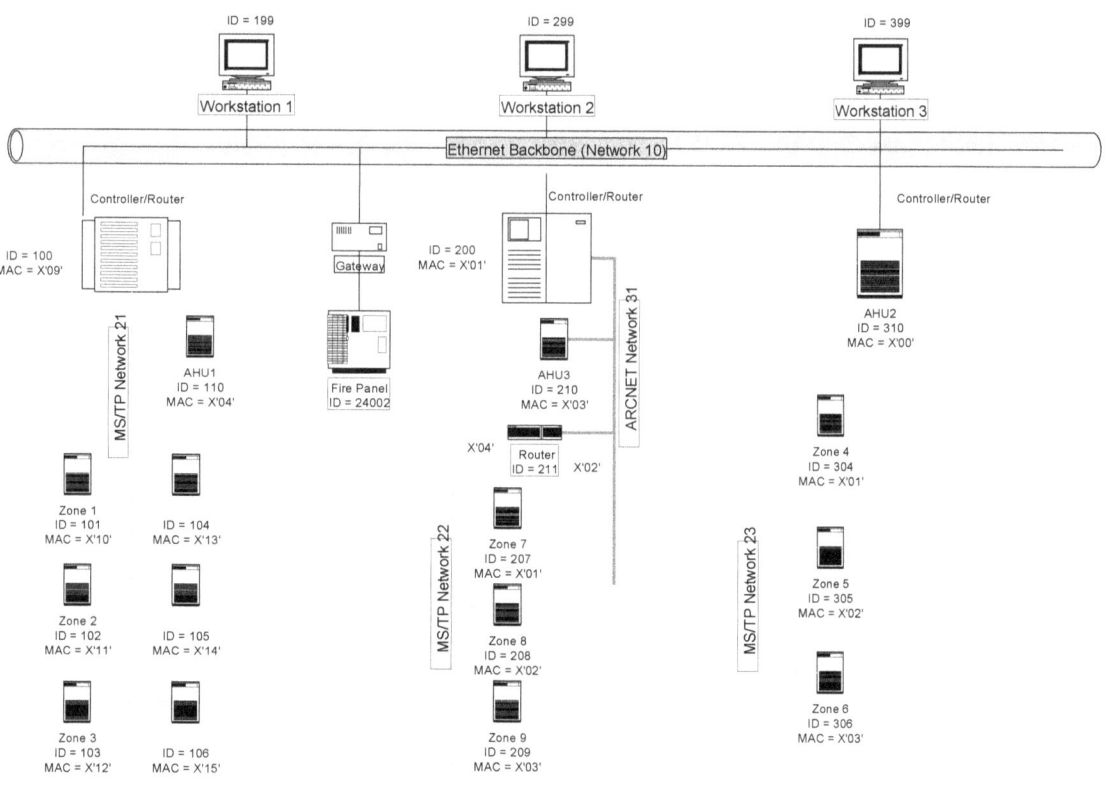

Figure 7. Network topology of the BACnet building automation and control system.

Products from three different control companies are used for HVAC controls. It is possible to monitor essential input and output values of all products from controller workstations using BACnet.

Air-handling unit (AHU) controllers in the virtual building provide supply fan control, return fan control, supply air temperature control, enthalpy economizer control, and supply air temperature reset. Appendix A provides an example BACnet object database for the controller for AHU 1. AI, AO, AV, BI, BO, and BV denote analog input, analog output, analog value, binary (digital) input, binary output, and binary value objects respectively. There is some variation in the BACnet object databases and control implementation details of the BACnet components. These differences reflect the fact that product features vary from manufacturer to manufacturer. The ranges of sensors and analog values in Appendix A correspond to design assumptions about the building equipment operating conditions. These can be varied by AHU or by zone.

Control of VAV boxes in the virtual building involves damper and reheat valve controls. The damper is controlled to modulate supply airflow to a zone, while the reheat valve coupled with the reheat coil enables the temperature of air supplied to the zone to be increased. Appendix B provides an example BACnet object database for a sample VAV box controller.

Appendices A and B illustrate the kind of detailed information that is available through a BACnet interface to the VCBT. They also provide the object identifier details needed to access the data using BACnet messages and the scaling details of the connection to the data acquisition system. Proportional-integral control is used for most local loops.

Some AHU and VAV box control strategies implemented in controllers are briefly described below.

Supply Fan Control

A typical, pressure-independent VAV system maintains a constant pressure at the VAV box inlets by sensing and controlling the static pressure at the supply duct. The supply fan controller with a proportional-integral (PI) algorithm receives the static pressure sensor output and sends the control output to a variable frequency motor controller to vary the supply fan speed.

The PI control is a feedback control to minimize the error signal $e(t)$.

$$e(t) = SP - MV(t)$$

where SP = set point
  $MV(t)$ = measured value
  t = time

A popular PI algorithm is expressed [17] as:

$$u(t) = u_0 + K_c [e(t) + \frac{1}{\tau_I} \int_0^t e(t^*) dt^*]$$

where $u(t)$ = controller output
u_0 = bias value
K_c = proportional gain
τ_I = integral time or reset time
t^* = dummy variable of integration

Return Fan Control

The airflow rates of the supply and the return systems are measured at the airflow stations. The return fan speed is controlled to maintain a constant flow rate difference between the supply airflow rate and the return airflow rate. A PI algorithm is employed.

Supply Air Temperature Control

The supply air temperature controller receives the sensor output from the supply air temperature sensor, and determines the appropriate positions of the cooling coil valve, heating coil valve, and mixing box dampers. It is assumed that outside air, exhaust air, and re-circulating air dampers are linked together to be controlled by a single controller output signal. Whenever the supply fan is on, the cooling and heating coil valves operate in conjunction with mixing box air dampers to maintain the set point value of supply air temperature. A PI controller is used.

Enthalpy Economizer Control

The sensor outputs from the outside air temperature, outside air relative humidity, return air temperature, and return air relative humidity sensors allow the enthalpy economizer controller to determine the damper openings. When the outside air enthalpy is greater than the return air enthalpy, the outside air damper is closed to its minimum position, and the cooling coil provides the required cooling. When the outside air enthalpy is less than the return air enthalpy and the outside air temperature is greater than the supply air temperature, the outside air damper is fully open, and the cooling coil supplements the required cooling.

Supply Air Temperature Reset

Based upon data from individual zone VAV boxes, the set point for supply air temperature can be reset to satisfy the maximum cooling demand required each zone during cooling season.

VAV terminal box controller

The airflow rate of a VAV terminal unit is governed by the damper opening. The reheat coil provides heating. The amount of heat is determined by the reheat coil valve opening. Typically the VAV terminal box controller controls the VAV box with a reheat coil in the way as shown in Figure 8. The zone temperature, T_z, is to be maintained near the set points ($T_{cool,set}$ for cooling and $T_{heat,set}$ for heating).

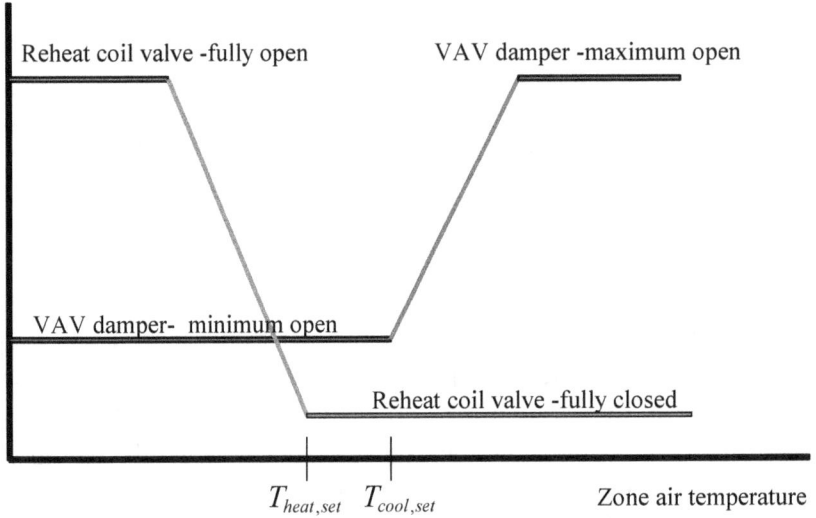

Figure 8. VAV box control

When $T_z \geq T_{cool,set}$, the damper opening modulates linearly until the maximum opening is reached.
When $T_z < T_{cool,set}$, the damper opening is at the minimum.
When $T_z \geq T_{heat,set}$, the reheat coil valve is fully closed.
When $T_z < T_{heat,set}$, the reheat coil valve modulates linearly until the maximum opening is reached.

2.6 BACnet Fire Panel

The old BACnet fire panel has been replaced with a new advanced fire panel. The previous fire panel could monitor addressable smoke detectors that were triggered by on/off operation. The new product can obtain analog values in voltage from addressable smoke detectors and heat sensors. First responders can monitor the information in the fire panel remotely using BACnet via Ethernet. In the virtual cybernetic building emulation, zone temperatures and zone smoke concentrations are be generated by ZFM-HVAC, transmitted to the data acquisition system, converted into analog data in voltage, sent to sensors, and sent to the BACnet fire panel as shown in Figure 9 where T and V denote the temperature and the voltage, respectively. When a zone/zones condition is in an alarm state due to fire, the fire panel provides alarms.

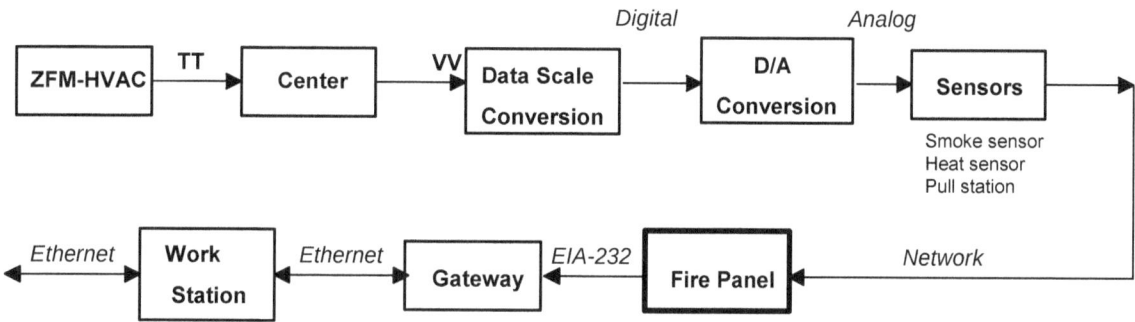

Figure 9. BACnet fire panel and ZFM-HVAC

2.7 Biometric Devices for Security

Biometric devices for eye recognition and hand recognition as security measures for door entry have been added to the VCBT [18]. All implemented devices are fully compatible with the BACnet protocol.

2.8 Sensor-Driven Fire Model (SDFM)

The sensor-driven fire is described elsewhere [12].

3. ZFM-HVAC

In the previous version of the VCBT, fire simulation was performed using the Consolidated Model of Fire Growth and Smoke Transport (CFAST) [19], and building HVAC system simulation was done using the HVACSIM$^+$ program separately. Communication between CFAST and HVACSIM$^+$ was accomplished by exchanging data through the Center using CORBA. CFAST and HVACSIM$^+$ had their own independent equation solving routines.

The new program called ZFM-HVAC is a computer program that combines a newly developed zone fire model, ZFM, and the building HVAC components based on HVACSIM$^+$.

The primary purpose of developing the combination program, ZFM-HVAC, is to organize all the computations under a single solver to ensure convergence to the proper solution. This consolidation will also aid in making the simulation expandable to large buildings.

3.1 Zone Fire Model (ZFM)

The zone fire model is described elsewhere. An early version can be found in reference [12].

3.2 HVAC Components from HVACSIM$^+$

A building HVAC system delivers conditioned air to building zones. The virtual HVAC system determines pressures, flow rates, temperatures and humidity values, while the controllers provide control signals. As mentioned previously, HVACSIM$^+$ solves a system of non-linear equations simultaneously; however HVAC-side equations of the ZFM-HVAC are solved sequentially. This sequential solving method was chosen, because there are many variables to be solved in the HVAC-side. The ZFM-side equation solver solves simultaneously only for zone/room-related variables such as layer temperature, pressure, density, and combustion product concentrations. Fire is modeled as a two-layer environment with a hot upper smoke layer and a cool lower layer. The time scale for these zone/room variables to change is short compared to the response of an HVAC system. The variables determined by HVAC-side are treated as pseudo-steady state variables due to the fact that most HVAC motor-driven actuators respond very slowly to control signals and final element components such as coils, and dampers are connected to the motor-driven actuators.

Most component models used in the HVAC-side of ZFM-HVAC originated from ASHRAE 825-RP by Norford and Haves [14] as "TYPE" routines. Brief descriptions of the models will be made based upon the report and the computer source code. Figure 10 shows a number of routines involved in the HVAC-side of ZFM-HVAC. Slightly modified HVAC component models of HVACSIM$^+$ are used except that the mixing box component model is greatly simplified and smoke mixing is added. The routine for merge flow (Type422) was also modified to add smoke mixing. Completely new routines for determining calculation sequences were developed solely for ZFM-HVAC using Fortran 90. At present, input data structures similar to HVACSIM$^+$ are used. The equations that are valid within their operating limits are presented below. Beyond the limits, the equations may not be valid. Special treatments are provided in the computer code for the equations in invalid regions.

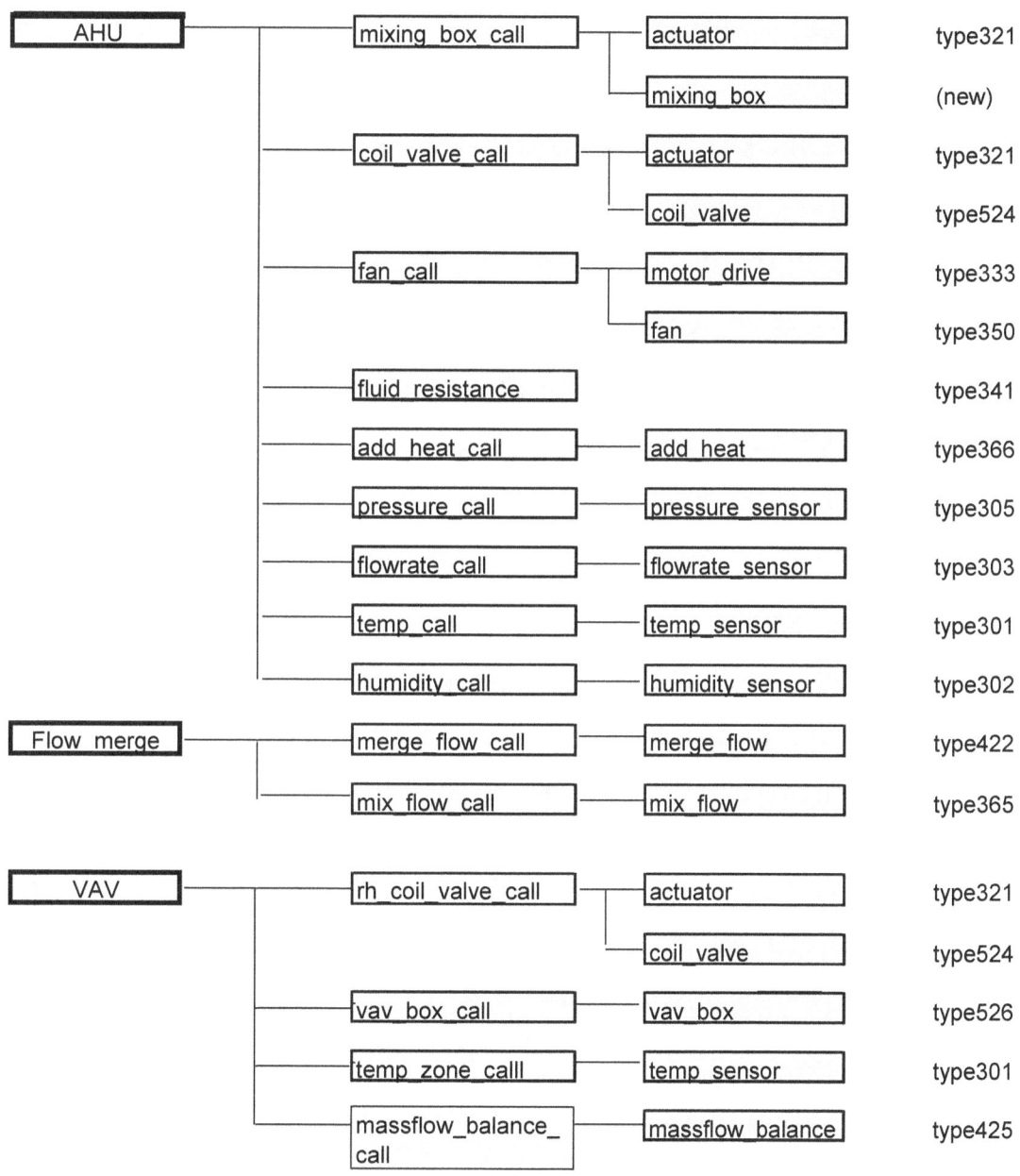

Figure 10. Routines involved in the HVAC-side of ZFM-HVAC.

3.2.1 Actuator (Type321)

The motor-driven actuator model calculates the position of a constant speed, motor-driven actuator accounting for hysteresis, crank geometry and range mismatch. Detailed description can be found in reference [14].

At each time step, the dimensionless actuator drive shaft position, $\theta(t)$, moves to next position according to the demand position, $\theta_D(t)$, and the state of movement of the motor. The possible states of the motor are backward motion ($\delta = -1$), stationary ($\delta = 0$), or forward motion ($\delta = +1$).

The maximum change of position during one time step, $\Delta\theta_{max}$, is:

$$\Delta\theta_{max} = \frac{\Delta t}{\tau_f}$$

where Δt is the duration of time step, and τ_f is the time required for the actuator to travel its full range from $\theta = 0$ to $\theta = 1$.

Three cases can be considered:

If the actuator motor was stationary at the previous time step ($\delta(t-1) = 0$), the new position remains at the previous position ($\theta(t) = \theta(t-1)$) and no movement occurs ($\delta(t) = 0$).

If the actuator motor was moving either forward or backward ($\delta(t-1) = +1$ or $\delta(t-1) = -1$), and $|\theta_D(t-1) - \theta(t-1)| \leq \Delta\theta_{max}$, the actuator moves to new position as demanded at the previous time step ($\theta(t) = \theta_D(t-1)$) and the actuator motor stops ($\delta(t) = 0$), if the demand position is attained.

If the actuator motor was moving ($\delta(t-1) = \pm 1$), but it could not reach the demanded position during the current time step ($|\theta_D(t-1) - \theta(t-1)| > \Delta\theta_{max}$), the motor continues to move by $\Delta\theta_{max}$. The new position is

$$\theta(t) = \theta(t-1) + \Delta\theta_{max}, \text{ if } \theta_D(t-1) \geq \theta(t-1)$$
$$\theta(t) = \theta(t-1) - \Delta\theta_{max}, \text{ if } \theta_D(t-1) < \theta(t-1)$$

The motor moving direction at the end of previous time, $\delta_e(t-1)$, becomes

$$\delta_e(t-1) = 1, \text{ if } \theta_D(t-1) \geq \theta(t-1)$$

1    $\delta_e(t-1) = -$,   if $\theta_D(t-1) < \theta(t-1)$

The new motor moving direction at the current time, $\delta(t)$, is determined considering whether the motor was moving in the direction required to reduce the position error at the end of the previous time step. If the motor was in the desired state at the end of the previous time step, no change of state is made ($\delta(t) = \delta_e(t-1)$).

The rotational motion of a motor-driven actuator matches the rotation motion of a damper, but most of valves need translational motion of the valve stem either by means of a screw thread or by a crank. When a valve is driven by a screw, the relationship between the dimensionless rotation position, $\theta(t)$, and the dimensionless translational position, $y(t)$, is linear. Thus, simply

) $y(t) = \theta(t)$

However, if a crank on the drive shaft is used, non-linearity needs to be considered. Figure 11 illustrates a motor-driven actuator with a crank on it along with a flow valve and a linkage. The length of the crank, R, the angular range of the crank, α, the total vertical displacement due to the cranking, Y, and the length of linkage, L, are given. The vertical displacement of the crank, $\tilde{y}$, corresponding to the angular position, $\tilde{\theta}$, is determined ignoring any bending of the linkage under the assumption that $L \gg R$.

From the figure,

) $Y = 2R\sin(\alpha/2)$
) $\tilde{y} = R\sin(\alpha/2) + R\sin(\tilde{\theta} - \alpha/2)$

Defining $\theta = \dfrac{\tilde{\theta}}{\alpha}$ and $y = \dfrac{\tilde{y}}{Y}$, the relationship of $\theta(t)$ and $y(t)$ is determined.

$$y(t) = \frac{1}{2} + \frac{\sin(\alpha(\theta(t) - \tfrac{1}{2}))}{2\sin(\alpha/2)}$$

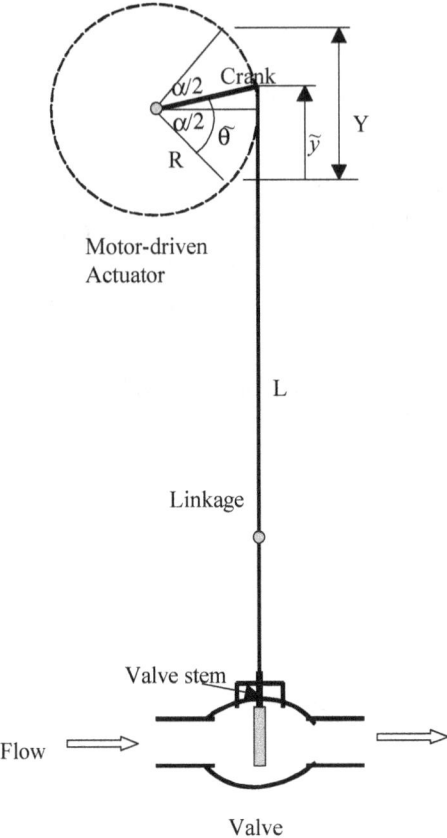

Figure 11. A crank of motor-driven actuator, a linkage, and a flow valve.

3.2.2 Mixing box

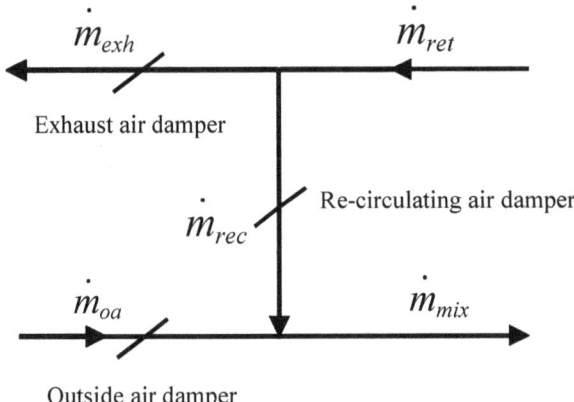

Figure 12. An air-handling unit mixing box

The mixing box of an air-handling unit consists of an outdoor damper, an exhaust damper, and a re-circulating damper as shown in Figure 4. Only one damper actuator is employed to control three dampers in responding to the PI control signal from the BMS

While the exhaust and outside air dampers are closing/opening in phase, the re-circulating damper is opening/closing.

This mixing box model used in ZFM-HVAC is a simplified model that differs from the model used in HVACSIM⁺.

The return (or extracted) air mass flow rate, $\dot{m}_{ret}$, is assumed to be

$$\dot{m}_{ret} = (1-\beta)\dot{m}_{mix}$$

where

$\dot{m}_{mix}$ = mixed air mass flow rate from the mixing box
β = fractional factor of air flow leakage in the air delivery system

Then the exhaust air mass flow rate, $\dot{m}_{exh}$, and the mixed air mass flow rate are expressed by

$$\dot{m}_{exh} = \dot{m}_{ret} - \dot{m}_{rec}$$
$$\dot{m}_{mix} = \dot{m}_{oa} + \dot{m}_{rec}$$

where the subscriptions *rec* and *oa* represent re-circulation and outside, respectively.

Assuming that the outside air mass flow rate is proportional to the outside air damper fractional opening, α,

$$\dot{m}_{oa} = \alpha \dot{m}_{mix}$$
$$\dot{m}_{exh} = (\alpha - \beta)\dot{m}_{mix}$$

The mixed air temperature, T_{mix}, becomes

$$T_{mix} = \frac{c_{p,oa}\dot{m}_{oa}T_{oa} + c_{p,rec}\dot{m}_{rec}T_{rec}}{c_{p,mix}(\dot{m}_{oa} + \dot{m}_{rec})}$$

where c_p is moist air specific heat.

The smoke concentration in the mixed air flow, C_{mix}, can be computed in terms of the smoke concentration of the return air flow, C_{ret}, since the outside air is smoke-free.

$$C_{mix} = C_{ret}(1-\alpha)$$

3.2.3 Temperature Sensor (Type301)

Temperature sensor is modeled by a first-order differential equation with a single time constant, τ_t.

The sensor output signal, c_o is give in terms of modified temperature input, c_i

$$c_i = \frac{T_i - T_{offset}}{G_t}$$

$$\tau_t \frac{dc_o}{dt} + c_o = c_i$$

where

T_i = temperature input
T_{offset} = offset for the minimum allowable temperature
G_t = temperature sensor gain

The differential equation is solved analytically at each time step, Δt.

$$c_o(t) = [c_o(t-1) - c_i]e^{-\Delta t/\tau} + c_i$$

The output sensor signal is bound between the maximum and minimum values. For each air-handling unit, there are four temperature sensors for the supply air, return air, mixed air, and outside air temperatures. These sensor outputs are inputs to the air-handling unit controller. One temperature sensor is also used in the return air duct of each zone to provide the input signal to the VAV controller for the zone.

3.2.4 Humidity Sensor (Type302)

Humidity sensor is modeled by a first-order differential equation similar to the temperature sensor model.

The sensor output signal, x_o, is give in terms of modified humidity input, x_i.

$$x_i = \frac{h_i - h_{offset}}{G_h}$$

$$\tau_h \frac{dx_o}{dt} + x_o = x_i$$

where

h_i = humidity ratio
h_{offset} = offset

G_h = humidity sensor gain
τ_h = humidity sensor time constant

There are two humidity sensors for the return air and outside air for each air-handling unit.

3.2.5 Flow rate Sensor (Type303)

A first-order differential equation is used to model the flow rate sensor.
The sensor output signal, F_o, is give in terms of modified flow rate input, F_i.

$$F_i = \frac{\dot{m}_i - \dot{m}_{offset}}{G_m}$$

$$\tau_m \frac{dF_o}{dt} + F_o = F_i$$

where
$\dot{m}_i$ = mass flow rate input
$\dot{m}_{offset}$ = offset
G_m = flow rate sensor gain
τ_m = flow rate sensor time constant

There are two flow rate sensors for the supply air and return air for each air-handling unit.

3.2.6 Static pressure Sensor (Type305)

The static pressure sensor output signal, P_o, is give in terms of modified static pressure input, P_i.

$$P_i = \frac{p_i - p_{offset}}{G_p}$$

$$\tau_p \frac{dP_o}{dt} + P_o = P_i$$

where
p_i = static pressure input
p_{offset} = offset
G_p = pressure sensor gain

τ_p = pressure sensor time constant

Only one static pressure sensor for the supply air is employed for each air-handling unit.

3.2.7 Motor drive (Type333)

The motor drive receives a demanded fractional speed signal, ω_d, from the BMS to determine the actual rotational speed, Ω_a, of an ideal motor of a fan. Motor dynamics, drive hysteresis, and non-linearity are neglected. Power consumption or efficiency is not considered.

The actual fractional speed, $\omega_a(t)$, is determined depending on the difference between current fractional demanded speed, $\omega_d(t)$, and previous actual fractional speed, $\omega_a(t-1)$.

$$\omega_a(t) = \omega_a(t-1) + \Delta\omega_{max}, \text{ if } \omega_d(t) \geq \omega_a(t-1)$$

$$\omega_a(t) = \omega_a(t-1) - \Delta\omega_{max}, \text{ if } \omega_d(t) < \omega_a(t-1)$$

where $\Delta\omega_{max}$ is the maximum change in fractional speed in one time step given by

$$\Delta\omega_{max} = \frac{\Delta t}{T_{trav}}$$

The travel time, T_{trav}, is the time required for the motor drive to travel from lower to upper limit.

With a given maximum rotational speed, Ω_{max}, the actual rotational speed is determined.

$$\Omega_a(t) = \omega_a(t)\Omega_{max}$$

At each air-handling unit, two motor drives are connected to the supply fan and the return fan, respectively.

3.2.8 Fan (Type350)

The fan model calculates a pressure rise, Δp, in terms of the dimensionless flow, C_f, defined by

$$C_f = \frac{\dot{m}}{\rho \Omega D^3}$$

where

$\dot{m}$ = mass flow rate
ρ = density of air
Ω = rotational speed
D = diameter of fan blade

The dimensionless pressure head function, f_h, is obtained from dimensionless fan performance curves, and can be represented by the 4th order polynomials.

$$f_h(C_f) = a_0 + a_1 C_f + a_2 C_f^2 + a_3 C_f^3 + a_4 C_f^4$$

Coefficients, a_i, are determined empirically. In the valid region of the performance curves, the pressure rise is expressed by

$$\Delta p = 0.001 \rho \, \Omega^2 D^2 f_h$$

Fan efficiency, η, is also represented by the 4th order polynomial equation as

$$\eta = e_0 + e_1 C_f + e_2 C_f^2 + e_3 C_f^3 + e_4 C_f^4$$

Coefficients, e_i, are empirically determined. Within the range of validity of the efficiency polynomial, fan power consumption, E_{fan}, is determined.

$$E_{fan} = \frac{\dot{m} \Delta p}{\rho \eta}$$

3.2.9 VAV Box (Type526)

A typical VAV box with a reheat coil is shown in Figure 5. This VAV box model represents a pressure-independent VAV terminal box. The reheat coil is excluded here, but the actuator is included. Only the zone supply airflow rate variation due to the VAV damper opening is considered. A VAV terminal box controller sends a signal of demand flow rate to a VAV box. In responding to the signal, the VAV damper opening varies.

For a given size of VAV box as nominal volumetric flow rate, V_{nom}, fractional demand speed of actuator motor, ω_d, is determined.

$$\omega_d = K_p (V_d - V_m), \quad \omega_{d,min} \leq \omega_d \leq 1.0$$

where
    K_p = proportional gain
    V_d = fractional demand volumetric flow rate
    $\omega_{d,min}$ = minimum fractional speed of motor

The measured fractional volumetric flow rate at the previous time step, $V_m(t-1)$, is

$$V_m(t-1) = \frac{\dot{m}(t-1)}{\rho V_{nom}}$$

Taking the average fractional demand speed during the time step, $\overline{\omega}_d$, new angular position of the actuator, $\theta(t)$, is

$$\theta(t) = \theta(t-1) + \overline{\omega}_d \frac{\Delta t}{T_{trav}}, \quad 0 \leq \theta(t) \leq 90°$$

The travel time, T_{trav}, is the time required for the motor drive to travel from 0° to 90°. In turbulent flow, the mass flow rate through the VAV box is given in terms of the pressure drop, Δp, and the total flow resistance, r_{tot}.

$$\dot{m}(t) = \sqrt{\frac{\Delta p}{r_{tot}}}$$

The total resistance is a sum of fixed system resistance, r_{fix}, and damper position-dependent resistance, r_θ.

$$r_{fix} = \frac{\Delta p_{open}}{\rho^2 V_{nom}^2} - \frac{0.001 k_{open}}{2\rho A_f^2}$$

where
  Δp_{open} = pressure drop at normal flow with fully open damper
  k_{open} = loss coefficient of fully open damper
  A_f = face area of damper

Legg's correlation is used to calculate the damper position-dependent resistance.

$$\ln(k_\theta) = \alpha + \beta\theta(t), \quad \begin{array}{l} 15° < \theta < 55° \text{ for opposed blade damper} \\ 15° < \theta < 65° \text{ for parallel blade damper} \end{array}$$

α and β are empirically determined constants. Using damper loss coefficient, k_θ, the damper resistance is represented as

$$r_\theta = \frac{e^{k_\theta}}{C}$$

where C is the conversion factor.

3.2.10 Fluid Resistance (Type341)

The flow resistance model is based on a square law relationship at high flow rates and a linear relationship at low flow rates. This routine calculates the inlet pressure, p_i, for given outlet pressure, p_o, mass flow rate, $\dot{m}$, and flow resistance, r.

$$p_i = p_o + r\dot{m}|\dot{m}| \quad \text{in turbulent flow}$$
$$p_i = p_o + r\dot{m}_{crit}\dot{m} \quad \text{in laminar flow}$$

where $\dot{m}_{crit}$ is the critical mass flow rate, which is the lower limit of purely quadratic flow.

3.2.11 Add Heat (Type366)

The outlet temperature of a fan or a duct, T_o, is calculated by adding the temperature rise, ΔT, due to added heat and the inlet temperature, T_i. ΔT is obtained analytically from

$$\tau_h \frac{d\Delta T}{dt} + \Delta T = \frac{\dot{Q}_{add}}{c_p \dot{m}}$$

where
$\dot{Q}_{add}$ = rate of heat addition
τ_h = time constant
c_p = specific heat of dry air

3.2.12 Merge Flow (Type422)

The outlet mass flow rate is the sum of the two inlet mass flow rates.

$$\dot{m}_o = \dot{m}_{i,1} + \dot{m}_{i,2}$$

The outlet smoke concentration, C_o, is calculated in a simple way.

$$C_o = \frac{C_{i,1}\dot{m}_{i,1} + C_{i,2}\dot{m}_{i,2}}{\dot{m}_o}$$

3.2.13 Mix Flow (Type365)

When multiple moist air flow streams merge, the outlet mixed air temperature, T_{mix} and the humidity ratio, h_{mix}, are obtained from

$$h_{mix} = \frac{\sum h_i \dot{m}_i}{\sum \dot{m}_i}, \qquad (i=1,n)$$

$$T_{mix} = \frac{\sum \dot{m}_i T_i (c_{pa} + c_{pg} h_i)}{(\sum \dot{m}_i)(c_{pa} + c_{pg} h_{mix})}, \qquad (i=1,n)$$

where c_{pa} and c_{pg} are the specific heat of dry air and the specific heat of water vapor. Only two flow streams are used for ZFM-HVAC.

3.2.14 Mass Flow Balance (Type425)

An assumption is made in ZFM-HVAC that no air leaks to outside from a zone, and the outlet mass flow rate from a zone, $\dot{m}_{outlet}$, is the same as the inlet mass flow rate, $\dot{m}_{inlet}$.

$$\dot{m}_{outlet} = \dot{m}_{inlet}$$

3.2.15 Coil & Valve (Type524)

The combined model of heating/cooling coil and its three-way control valve is the most complex routine among the component models used in HVACSIM$^+$. The temperature and humidity of the airflow through heating or cooling coil are governed by the three-way coil valve, which is connected to a coil valve actuator. The actuator operates in responding to the control signal from the BMS.

The coil model used is a finned-tube cross-flow heat-exchanger with both fluids unmixed. When the working fluid is chilled water, the coil acts as a cooling coil, whereas when the working fluid is hot water, the coil is a heating coil. On the air-side of the heat-exchanger, the inlet air dry bulb temperature, $T_{a,i}$, inlet air humidity ratio, $h_{a,i}$, and dry air mass flow rate, $\dot{m}_a$, are given. On the water-side of the heat-exchanger, the inlet water Temperature, $T_{w,i}$, inlet water pressure, $p_{w,i}$, and outlet water pressure, $p_{w,o}$, are given. With these input conditions, the outlet air dry bulb temperature, $T_{a,o}$, outlet air humidity ratio, $h_{a,o}$, outlet water temperature, $T_{w,o}$, primary circuit water mass flow rate, $\dot{m}_{w,prim}$, water mass flow rate through the coil, $\dot{m}_w$, and mixed return water temperature, $T_{w,ret}$, are determined. Using the outlet air pressure, $p_{a,o}$, the inlet air pressure, $p_{a,i}$, is also calculated. Figure 13 is a schematic diagram of the coil with external water circuits.

The equations in a general case are presented, not exactly following the calculation logic flow of the routine.

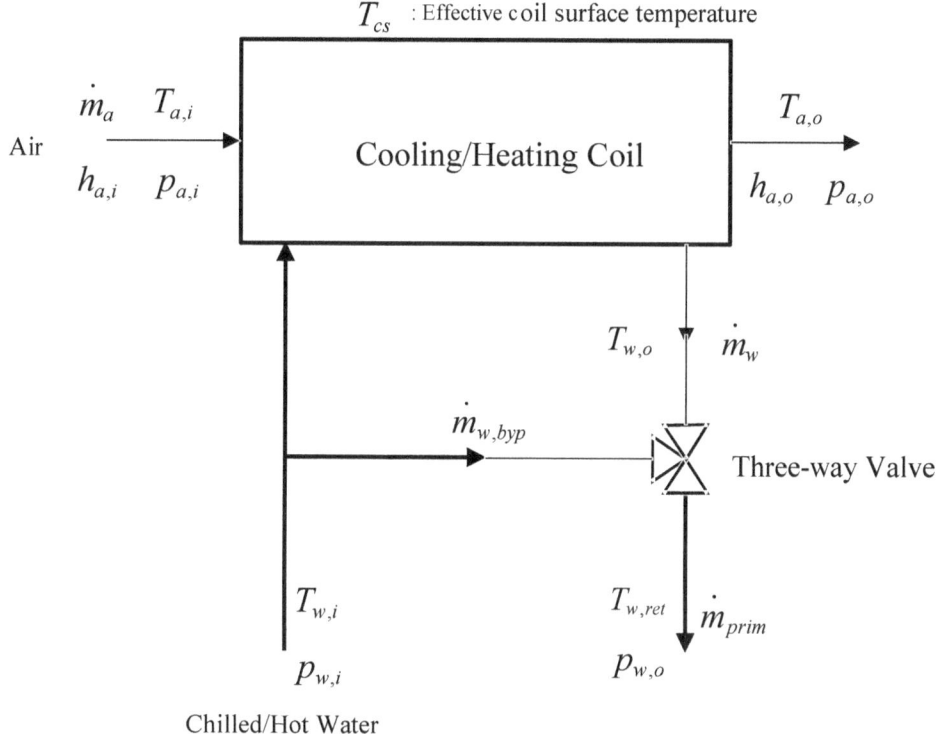

Figure 13. The cooling/heating coil and three-way valve.

Outlet air dry bulb temperature of a dry coil ($T_{a,o}$)

$$T_{a,o} = T_{a,i} + \frac{\dot{Q}_{a,dry}}{c_{pa}\dot{m}_a}$$

where

c_{pa} = specific heat of dry air

$\dot{Q}_{a,dry}$ = air-side heat transfer of a dry coil

$$\dot{Q}_{a,dry} = (T_{cs} - T_{a,i})C_a(1 - B_f)$$

$T_{c,s}$ = effective coil surface temperature

C_a = air capacity rate = $(c_{pa} + c_{pg} h_{a,i}) \dot{m}_a$

c_{pg} = water vapor specific heat

B_f = coil by-pass factor

Effective coil surface temperature (T_{cs})

The effective coil surface temperature is obtained by solving the first-order differential equation.

$$\frac{dT_{cs}}{dt} = \alpha T_{cs} + \beta$$

where

$$\alpha = \frac{r_1 + r_2}{r_1 r_2 HC_{coil}}$$

$$\beta = \frac{1}{HC_{coil}} \left(\frac{T_{a,i}}{r_1} + \frac{T_{w,i}}{r_2} \right)$$

HC_{coil} = heat capacity of coil = $HC_{fin} + HC_{tube} + HC_{water}$

$$r_1 = \frac{1}{C_a(1 - B_f)}$$

$$r_2 = \frac{1}{\varepsilon_{HX} C_{min}}$$

C_{min} = minimum heat capacity rate

ε_{HX} = heat-exchanger effectiveness

Coil by-pass factor (B_f)

$$B_f = e^{-\frac{A_{ext}}{C_a r_a}}$$

where

A_{ext} = heat-exchanger total heat transfer area on air-side

r_a = air-side thermal resistance = $\dfrac{1}{\tilde{h}_a \varepsilon_{fin}}$

$\tilde{h}_a$ = air-side heat transfer coefficient = $\dfrac{Nu_a k_a}{D_h}$

Nu_a = Nusselt number of air = $3.66 + \dfrac{0.0668\, Re_a\, Pr_a\, \dfrac{D_h}{L_{coil}}}{1 + 0.04\left(Re_a\, Pr_a\, \dfrac{D_h}{L_{coil}}\right)^{2/3}}$

Re_a = Reynolds number of air = $\dfrac{\dot{m}_a D_h}{A_{free} \nu_a}$

Pr_a = Prantl number of air = $\dfrac{c_{pa} \nu_a}{k_a}$

k_a = thermal conductivity of air
D_h = hydraulic diameter
L_{coil} = length of coil
A_{free} = heat-exchanger minimum free-flow area on air side
ν_a = dynamic viscosity of air = $\dfrac{1.458 \times 10^{-6}(273.16 + T_{a,i})^{1.5}}{383.4 + T_{a,i}}$

ε_{fin} = fin effectiveness

$$\varepsilon_{fin} = 1 - \left(\dfrac{A_{fin,net}}{A_{fin,net} + A_{tube,net}}\right)\left(1 - \dfrac{\tanh(\gamma)}{\gamma}\right)$$

$$\gamma = \sqrt{\dfrac{2\tilde{h}_a}{k_{fin} t_{fin}}}\left[1 + 0.35 \ln\left(\dfrac{R_{fin,o}}{R_{fin,i}}\right)\right] H_{fin}$$

$A_{fin,net}$ = net fin heat transfer area
$A_{tube,net}$ = net tube heat transfer area
k_{fin} = thermal conductivity of fin material
t_{fin} = fin thickness
$R_{fin,o}$ = effective fin radius
$R_{fin,i}$ = fin inside radius

Heat-exchanger effectiveness (ε_{HX})

The effectiveness of a compact cross-flow heat-exchanger with both fluids unmixed is calculated for a typical condition.

$$\varepsilon_{HX} = 1 - \exp\left[\frac{-NTU^{0.22}\left(1 - e^{-C_{rr}NTU^{0.78}}\right)}{C_{rr}}\right]$$

where

$$C_{rr} = \frac{C_w}{C_a} \quad \text{for water is minimum}$$

$$C_{rr} = \frac{C_a}{C_w} \quad \text{for air is minimum}$$

C_w = water heat capacity rate = $c_{pw}\dot{m}_w$

c_{pw} = specific heat of water

NTU = number of transfer units = $\dfrac{A_{ext}}{C_{min} r_{tot}}$

r_{tot} = sum of resistances of air-side, water-side, and material

Outlet air dry bulb temperature of a wet coil ($T_{a,o}$)

$$T_{a,o} = T_{a,i} + \dot{Q}_{a,wet} SHR$$

where

$$SHR = \text{sensible heat ratio} = \frac{C_a\left(T_{a,o,est} - T_{a,i}\right)}{\dot{Q}_{a,wet}}$$

$T_{a,o,est}$ = estimated outlet air dry bulb temperature

$$T_{a,o,est} = B_f\left(T_{a,i} - T_{cs}\right) + T_{cs}$$

$\dot{Q}_{a,wet}$ = air-side heat transfer of a wet coil

$$\dot{Q}_{a,wet} = (T_{cs} - T_{wb,i})C_{a,sat}(1 - B_f)$$

$T_{wb,i}$ = inlet air wet bulb temperature

$C_{a,sat}$ = saturated air capacity rate = $c_{p,sat}\dot{m}_a$

$c_{p,sat}$ = saturated air specific heat

Outlet air humidity ratio ($h_{a,o}$)

$$h_{a,o} = \frac{H_{a,o} - c_{pa}T_{a,o}}{c_{pg}T_{a,o} + HFG}$$

where

HFG = latent heat of vaporization

$H_{a,o}$ = outlet air enthalpy = $H_{a,i} + \dfrac{\dot{Q}_{a,wet}}{\dot{m}_a}$

Coil outlet water temperature ($T_{w,o}$)

$$T_{w,o} = T_{w,i} - \frac{\dot{Q}_w}{C_w}$$

where

$\dot{Q}_w$ = heat transfer on water-side = $\dfrac{T_{w,i} - T_{cs}}{r_2}$

Mixed return water temperature ($T_{w,ret}$)

$$T_{w,ret} = \frac{T_{w,i}(\dot{m}_{w,prim} - \dot{m}_w) + T_{w,o}\dot{m}_w}{\dot{m}_{w,prim}}$$

where

$\dot{m}_{w,prim}$ = primary circuit water mass flow rate

$\dot{m}_w$ = water mass flow rate through the coil

$T_{w,i}$ = inlet water temperature

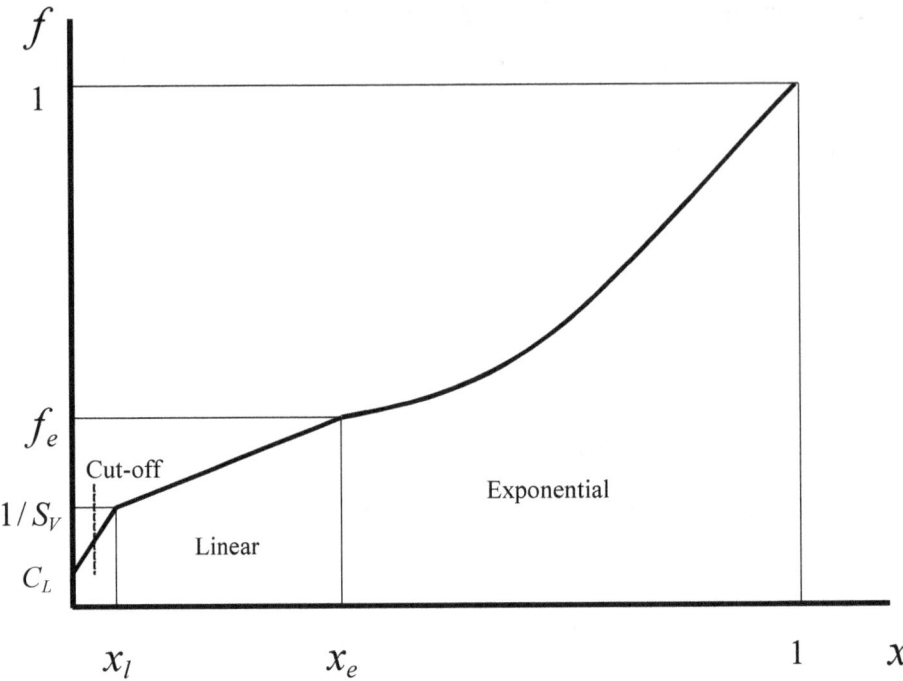

Figure 14. Water flow control valve characteristics

The water flow control valve characteristics used in the model in the general form is presented in Figure 14. The figure shows the relationship between the fractional flow rate through the control valve, $f(x)$, and the valve stem position, x. A constant pressure is applied. There are three regions – cut-off, linear, and exponential.

In the cut-off region,

$$f(x) = \left(\frac{1}{S_V} - C_L\right)\left(\frac{x}{x_l}\right) + C_L \qquad 0 \leq x \leq x_l$$

In the linear region,

$$f(x) = \left(\frac{f_e - \frac{1}{S_V}}{x_e - x_l}\right)(x - x_l) + \frac{1}{S_V} \qquad x_l < x \leq x_e$$

In the exponential region,

$$f(x) = a^{x-1} \qquad x_e < x \leq 1$$

where

S_V = rangeability – ratio of highest to lowest controllable flow
C_L = fractional leakage flow when the valve is fully closed
x_l = break point between the linear and cut-off regions
x_e = break point between the exponential and linear regions
a = curvature parameter for equal percentage characteristic
$f_e = f(x_e)$

The hydraulic resistance across the flow port of the control valve, $r_{w,v,f}$, is expressed in terms of the fractional flow rate and valve capacity, K_V.

$$r_{w,v,f} = \frac{1296}{K_V^2 f^2(x)}$$

The resistance across the bypass port of the valve, $r_{w,v,byp}$, is calculated from

$$r_{w,v,byp} = \frac{1296}{K_V^2 f^2(1-x)}$$

Water flow rate through the coil ($\dot{m}_w$)

$$\dot{m}_w = \sqrt{\frac{p_{w,i} - p_{w,o}}{r_{w,c} + r_{w,v,f}}}$$

where

$p_{w,i}$ = inlet water pressure
$p_{w,o}$ = outlet water pressure
$r_{w,c}$ = hydraulic resistance through the coil

By-pass water flow rate ($\dot{m}_{w,byp}$)

$$\dot{m}_{w,byp} = \sqrt{\frac{p_{w,i} - p_{w,o}}{r_{w,byp} + r_{w,v,byp}}}$$

where

$r_{w,byp}$ = hydraulic resistance through the by-pass water circuit

Primary water flow rate ($\dot{m}_{w,prim}$)

$$\dot{m}_{w,prim} = \dot{m}_w + \dot{m}_{w,byp}$$

Return water temperature ($T_{w,ret}$)

$$T_{w,ret} = \frac{T_{w,i}\dot{m}_{w,byp} + T_{w,o}\dot{m}_w}{\dot{m}_{w,prim}}$$

Inlet air pressure of the coil ($p_{a,i}$)

$$p_{a,i} = p_{a,o} + r_{a,c}\dot{m}_a|\dot{m}_a| \quad \text{above the critical bound}$$
$$p_{a,i} = p_{a,o} + r_{a,c}\dot{m}_{crit}\dot{m}_a \quad \text{below the critical bound}$$

where

$\dot{m}_{crit}$ = critical flow – lower limit of purely quadratic flow

Figure 15 is a diagram showing the airflows through various building HVAC component models used here. The four big blocks are AHU, VAV, Flow merge, and Fire Model. Air pressure, air mass flow rate, air temperature, air humidity ratio, and smoke concentration are denoted by p, m, t, h, and o, respectively. The numbers following the characters are indices chosen for identification. A component model is identified by "Ty" plus the specific number for the TYPE routine in HVACSIM$^+$ as seen in Figure 10. No specific number is given to the new mixing box routine in ZFM-HVAC. The same configuration is applied to all the three AHUs. Some simplification was made to the models compared with the model using HVACSIM$^+$. Particularly, a constant static pressure in the air supply duct is assumed during emulation.

As shown in the Figure 15, the fire model receives the supply air mass flow rate, supply temperature, supply pressure, and smoke concentration from HVAC duct as inputs to a zone. The fire model determines the zone outlet pressures, temperatures, and combustion product concentrations. No solar heat gain, nor internal heat gain in a zone is considered in normal status, which is no fire condition.

4. ZFM-HVAC RUNS AND DISCUSSION

Three scenarios were selected to demonstrate the use of ZFM-HVAC in the VCBT.

Case 1: A virtual three-story building with nine conditioned zones and an unconditioned stairwell. Three AHUs and nine VAV boxes are involved.

Case 2: A virtual single-story building with nine conditioned zones. Eight of nine zones are identically sized. Three AHUs and nine VAV boxes are used.

Case 3: A virtual single-story building with nine conditioned zones with different sizes. Three AHUs and nine VAV boxes are involved.

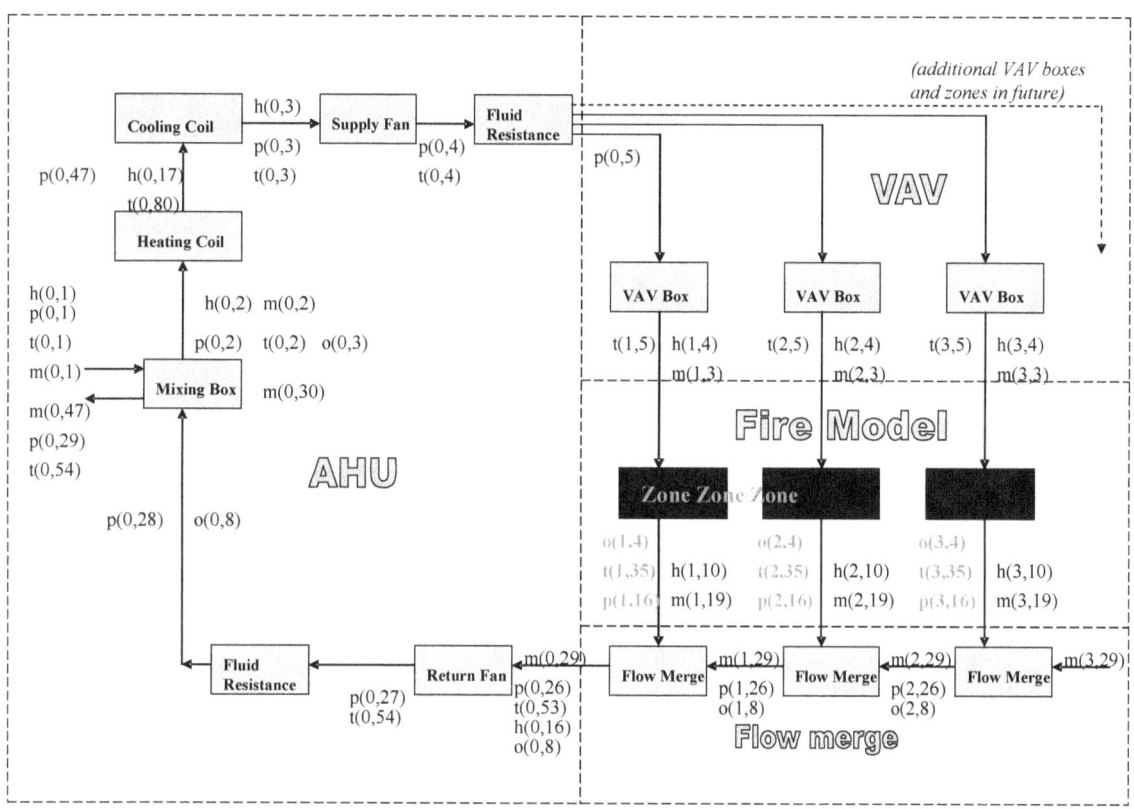

Figure 15. The airflow paths through various building HVAC components for all cases.

4.1 Case 1 Scenario

As an example run, a virtual three-story building as shown in Figure 16 was modeled. This virtual building has three zones on each floor and a stairwell. Figures 17 to 19 show the layouts for all three floors. As shown in Figure 20, three air-handling units serve three zones each. AHU1 serves Zones 1, 2, and 3; AHU2 serves Zones 4, 5, and 6; and AHU3 serves Zones 7, 8, and 9, respectively. None of the AHUs serves the stairwell (Zone 10). Access doors to the stairwell are located at Zone 7, 8, and 9. Two access doors to outside are located in Zone 7. Airflow is allowed between zones. The AHUs supply and extract the airflows through ducts and VAV boxes. It should be emphasized that the ZFM-HVAC was primarily designed to observe the effects on the zone air temperature and smoke concentration in the zones as well as HVAC system responses due to fire.

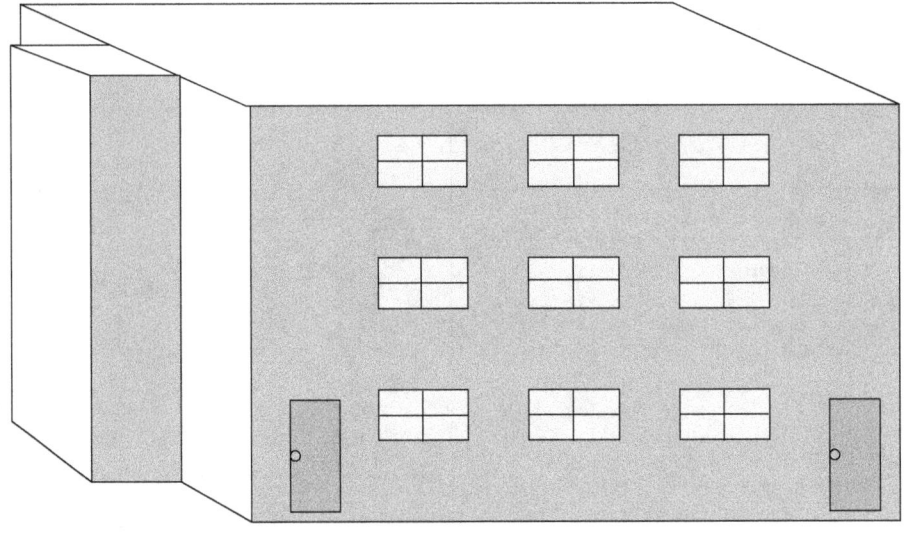

Not To Scale

Figure 16. A virtual three-story building in Case 1.

No controller tuning, FDD, nor HVAC system energy consumption analysis is emphasized with ZFM-HVAC. In all the cases, the supply and return fans were set to run continuously at all times.

Initially AHUs and VAV boxes operated in a normal mode in conjunction with the real BACnet HVAC controllers. Two different sizes of fire were used. The heat release rate determines the size of a fire. Figure 21 shows heat release rates in terms of time. These heat release rates were artificially imposed as input conditions. Case 1_1 and Case 1_2 represent the small fire and the large fire respectively. Appendix C is the ZFM-side input data for Case 1_1.

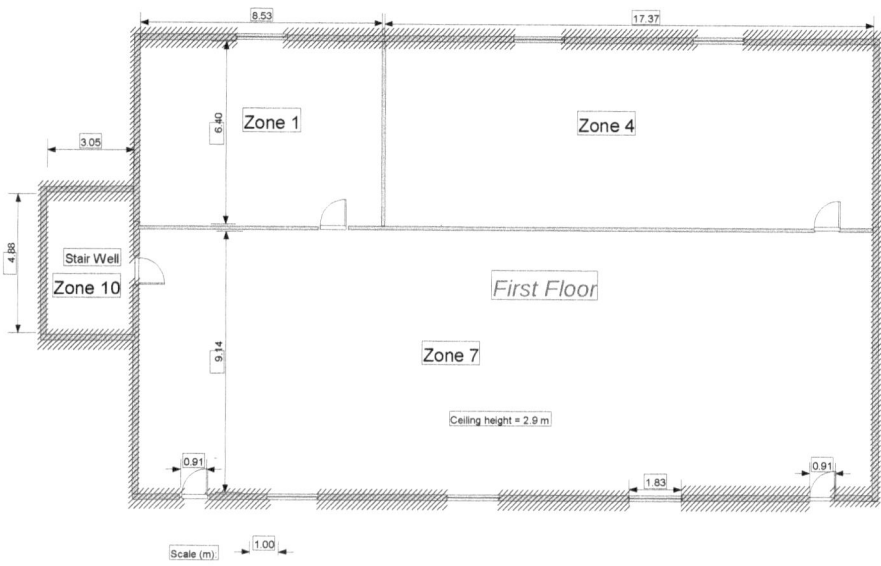

Figure 17. The first floor layout of the virtual three-story building in Case 1.

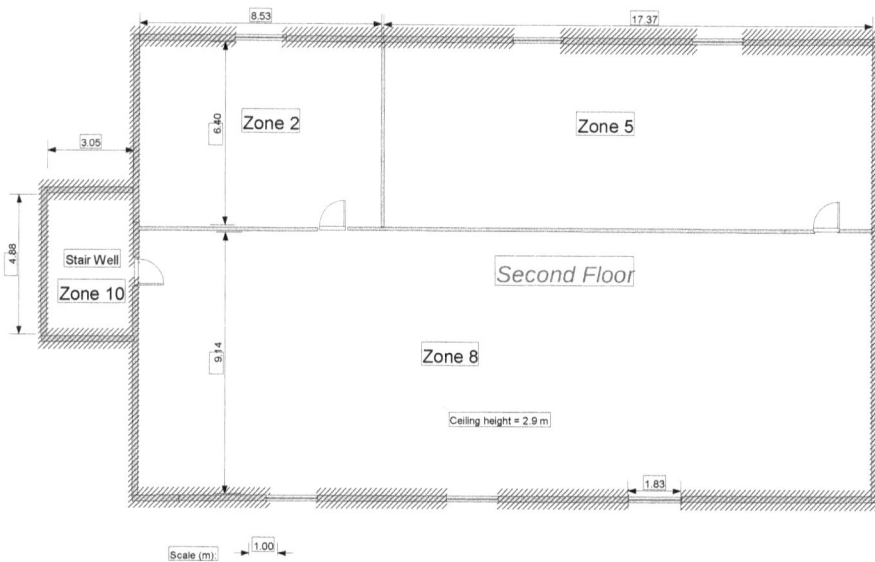

Figure 18. The second floor layout of the virtual three-story building in Case 1.

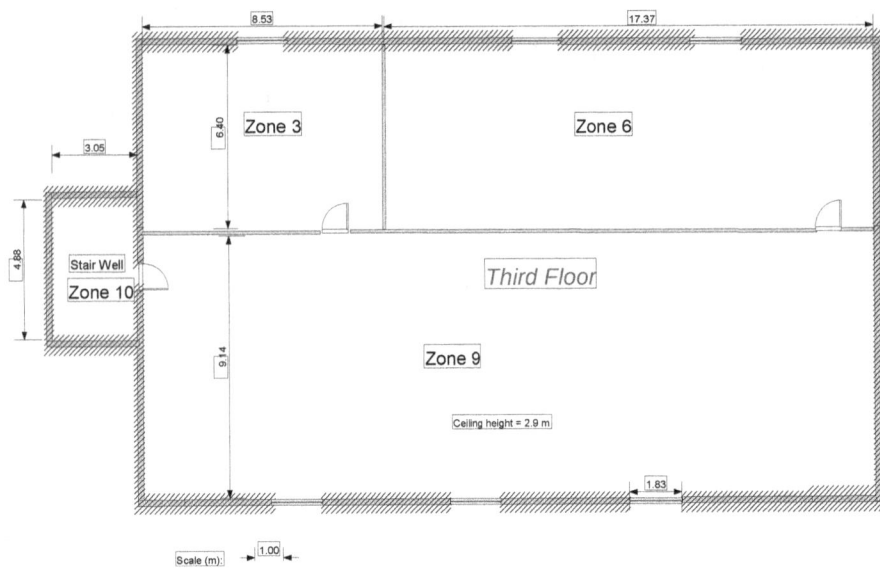

Figure 19. The third floor layouts of the virtual three-story building in Case 1.

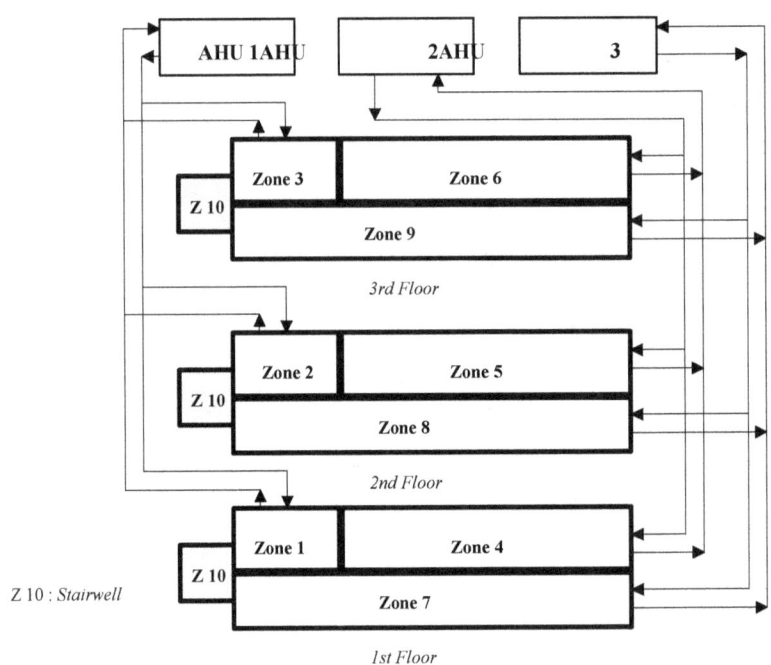

Figure 20. Air distribution for nine conditioned zones of the virtual three-story building in Case 1.

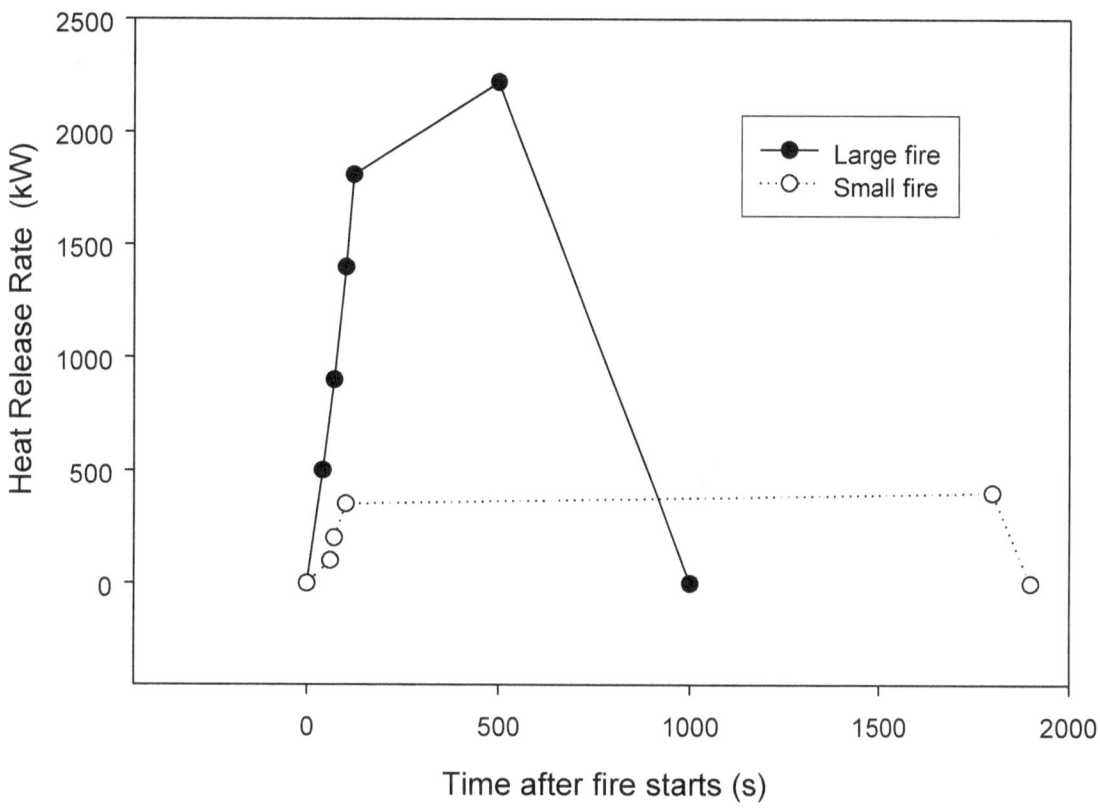

Figure 21. Heat release rates for large and small fires.

Appendix D shows the HVAC-side configuration input data, where n_ahu is the number of air-handling units and nbr is the number of flow branches for VAVs served by each air-handling unit. The files with extension 'inp' are input data files and those with 'out' are output files. Sample input data for AHU and VAV are given in Appendices E and F. The parameter data are actually used in computation. Others are for information. Appendix G lists initial data of the HVAC-side of ZFM-HVAC.

First, the VCBT emulation results of Case 1_1 for the small fire is examined. At 7200 s from the beginning of emulation, a fire was set in Zone 1. For the AHU1 branch, the BACnet controllers involved are the AHU controller (ID = 100), and the VAV controllers (ID = 101, 102, and 103). Refer to Figure 7. The control strategies actually implemented in these controllers are much complicated than the general strategies described previously (See Section 2.5). For example, the VAV controller sets an alarm condition and limits the supply flow rate at the minimum when the zone air temperature becomes above 50 °C, although generally we might expect the maximum rate instead. Note that emulation results of this report reflect the responses of controllers of different companies in different ways.

Figures 22 and 23 show the return air temperatures from zones, the air temperatures at AHU1, the flow rates of AHU1 branch, the damper opening demand by controllers of AHU1, the coil opening demand signals, and the reheat coil control signals. Referring to Figure 15, ZA_TMP_1 [t(1,35)], ZA_TMP_2 [t(2,35)], and ZA_TMP_3 [t(3,35)] represent the return air temperatures from Zone 1, 2, and 3, respectively. The bottom and top edges of rectangular return air ducts locate 1.5 m and 2.0 m from the floor respectively as shown in Appendix C. The fire side of the ZFM-HVAC program determines the return air temperatures. In this case, the changes in ZA_TMP_2 and ZA_TMP_3 are not noticeable, but ZA_TMP_1, which is the return air temperature from the fire zone, Zone 1, is very high as expected. The zone air temperatures on the 2nd and the 3rd floors are not much affected due to the small fire size.

The supply air temperature of AHU1, mixed air temperature, return air (extracted air) temperature, and outdoor air temperature are denoted as SA_TMP [t(0,4)], MA_TMP [t(t(0,2)], RA_TMP [t(0,54)], and OA_TMP [t(0,1)], respectively. Although SA_TMP was maintained around the set point, MA_TMP and RA_TMP approached 100°C during the fire. OA_TMP did not vary much in this emulation period. The flow rates in the flow branch served by AHU1 are plotted for SA_FLOW [m(0,2)], RA_FLOW [m(0,29)], ZFLOW_1 [m(1,3)], ZFLOW_2 [m(2,3)], and ZFLOW_3 [m(3,3)], which represent the supply airflow rate, return airflow rate, and the zone airflow rates of Zone1, 2, and 3, respectively.

Demand control signals from BMS were observed. The AHU damper opening demand signal (DMP_DEM), the cooling coil valve opening demand signal (CC_DEM), and the heating coil valve opening demand signal (HC_DEM) are shown in Figure 23a. The actual damper opening (DMP_opening) is also shown on the same chart. It can be seen that DMP_opening value remained the same value of 0.15, which means that approximately 15 % of the AHU fresh (outdoor) air damper, 15 % of the exhaust air damper, and 85 % of the AHU re-circulation damper were open. It is because the DMP_opening was set to the constant value of 15 % in ZFM-HVAC overriding DMP_DEM that varied between 0.15 and 1.0. The AHU1 controller operated with the enthalpy economizer logic enabled. The VAV box damper opening demand signals (ZDMP_CTL_1, ZDMP_CTL_2, ZDMP_CTL_3) and the reheat coil valve opening demand signals (RHC_CTL_1, RHC_CTL_2, RHC_CTL_3) are shown in Figure 23b.

(a) Temperatures at AHU1 and return air temperatures from zones 1, 2, and 3

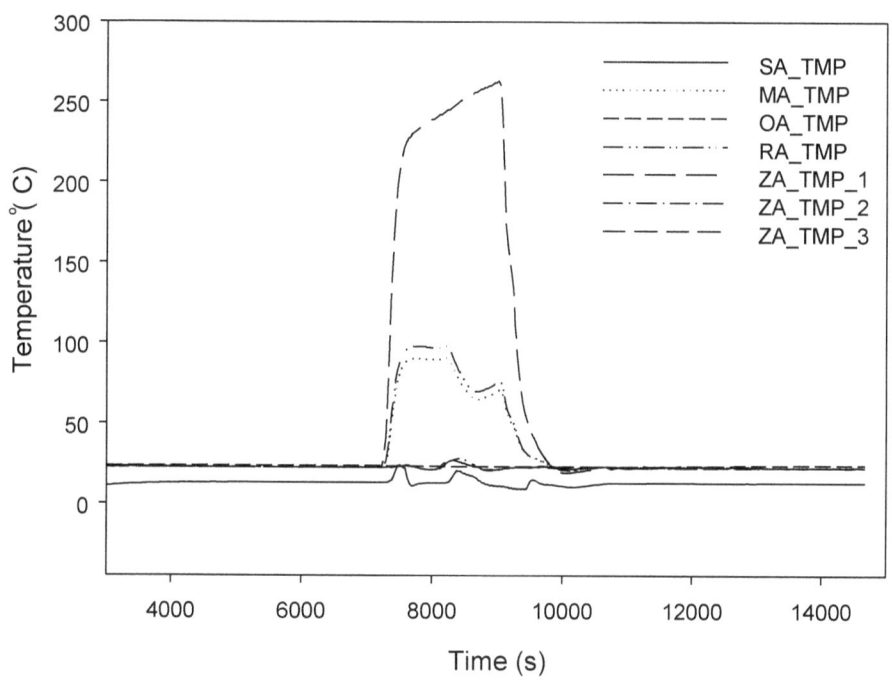

(b) Flow rates of AHU1 branch

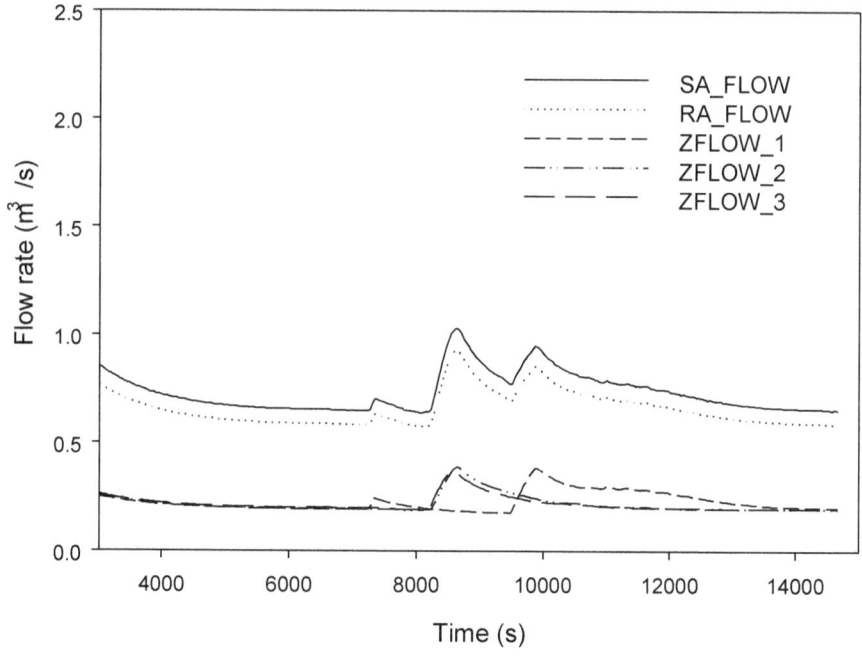

Figure 22. Temperatures, and flow rates of AHU1 in Case 1_1.

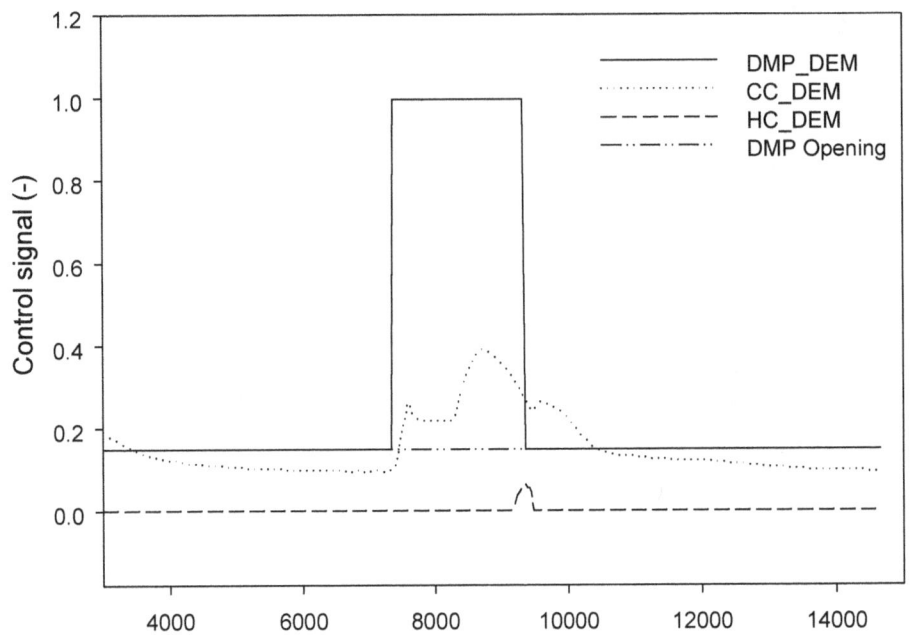

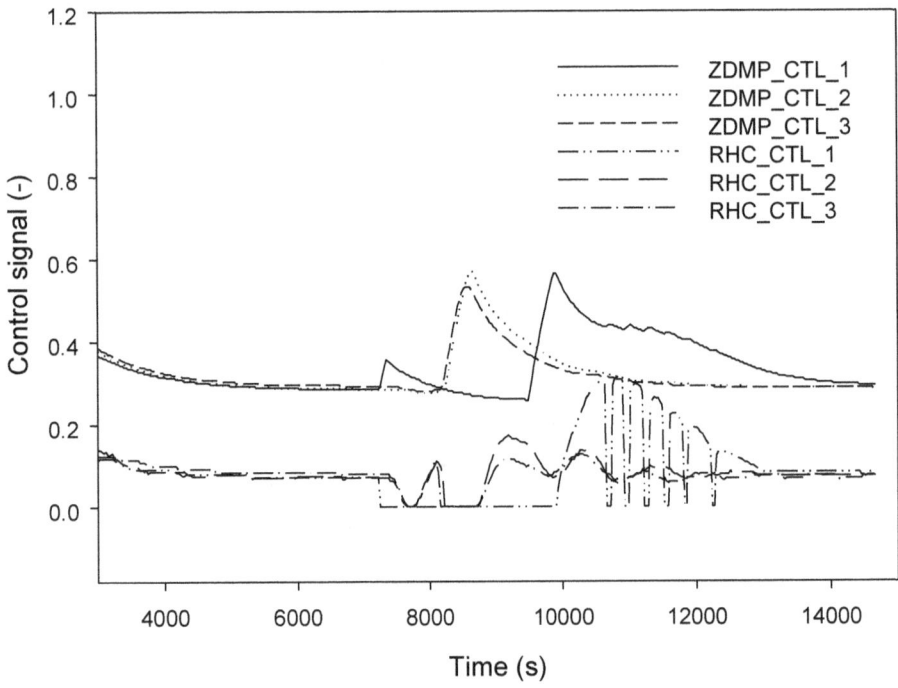

Figure 23. Control signals of AHU1 in Case 1_1. (-) means dimensionless.

In Case 1_2, the same building configuration as the previous case was applied, except that a large fire was set in the Zone 1 at 7,200 s after the start of emulation. Figure 24 shows the same kind of plots for AHU1 branch similar to Case 1_1. Notice that ZA_TMP_1 went up to 1600 °C, and ZA_TMP_2 and ZA_TMP_3 approached 160 °C. The airflow through the stairwell and the AHU ducts could contribute to the temperature increases. Since DMP opening was set to the minimum value of 15 %, roughly 85 % of the extracted flow from zones (RA_FLOW) returned to zones as the supply airflow after mixing with the fresh air. The supply air temperature became quite high during the period of fire. The cooling coil model in ZFM-HVAC was programmed up to 100 °C. Beyond the limit, the cooling coil discharge temperature was set to equal the return air temperature ensuring that no cooling would occur.

It is worthwhile to mention about why the reheat coils were activated by the VAV terminal box controller during the cooling season under no fire condition when the outdoor air temperature (OA_TMP) of the select weather condition was at 23.5 °C to 24.0 °C. The set point of the supply air temperature of the AHU was set at 12.8 °C, while the set point of the zone VAV controllers was 23.4 °C. No heat source (e.g., generated heat by person, equipment, or lighting, and heat transfer from ceiling, floor, or wall) in a zone was enabled. The only way the zone temperature could reach the set point value was reheating by the reheat coils. As a result, active reheat coil valve control signals can be observed.

Figures 26 and 27 show temperatures, flow rates, and control signals of the air-handling unit AHU2 and the VAVs served by AHU2 in Case 1_2. ZA_TMP_4, ZA_TMP_5, and ZA_TMP_6 represent the return air temperatures from Zone 4, 5, and 6, respectively. See Figures 17 through 20. For the AHU2 branch, the BACnet device instances of the AHU controller and the VAV controllers are 310, 304, 305 and 306 as shown in Figure 7. Compared with Figure 24, the return air temperatures from zones were much lower. The peak value was about 150 °C. The flow rates as well as control loops for the AHU2 branch quite fluctuated.

Figures 28 and 29 depict the similar plots as before for the AHU3 branch. Zones 7, 8, and 9 have openings to Zone 1, 2, and 3. This would bring the return air temperatures higher than those of AHU2. ZA_TMP_7 went up to 500 °C and ZA_TMP_8 and ZA_TMP_9 reach near 180 °C. Note that SA_TMP followed a similar pattern as MA_TMP or RA_TMP, which was near 260 °C at its peak. Here again as seen in Figure 29a, DMP_DEM was sometimes at 100 %, but the actual damper opening was set to 15 % at all times by ignoring the output signal, DMP_DEM, from the AHU enthalpy control.

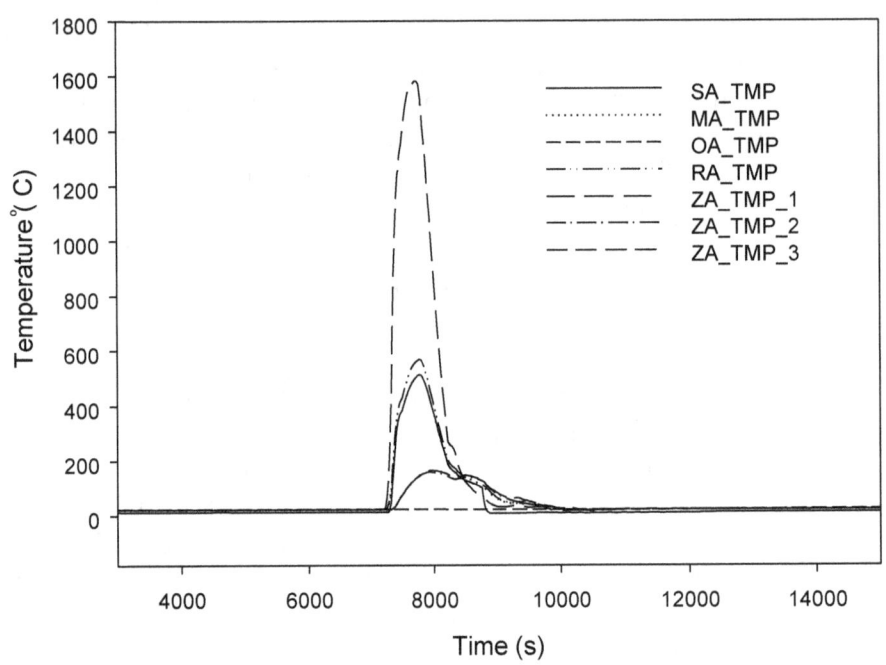

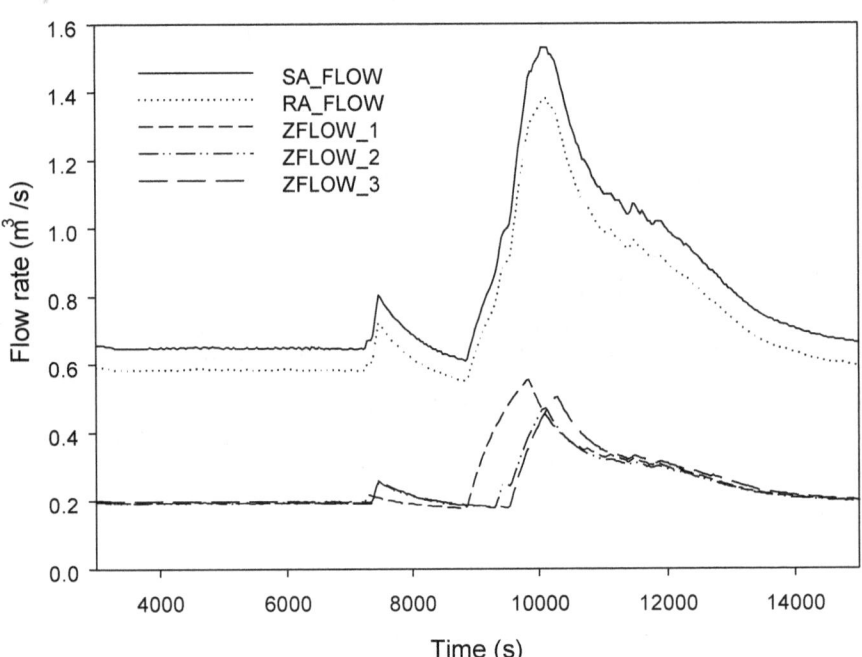

Figure 24. Temperatures, and flow rates of AHU1 in Case 1_2.

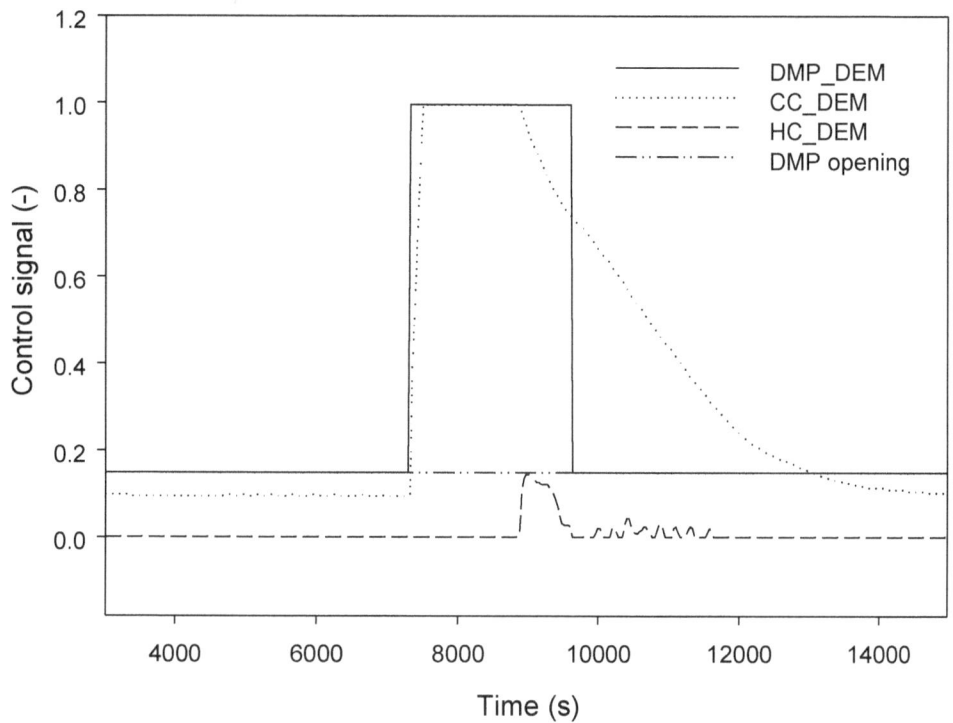

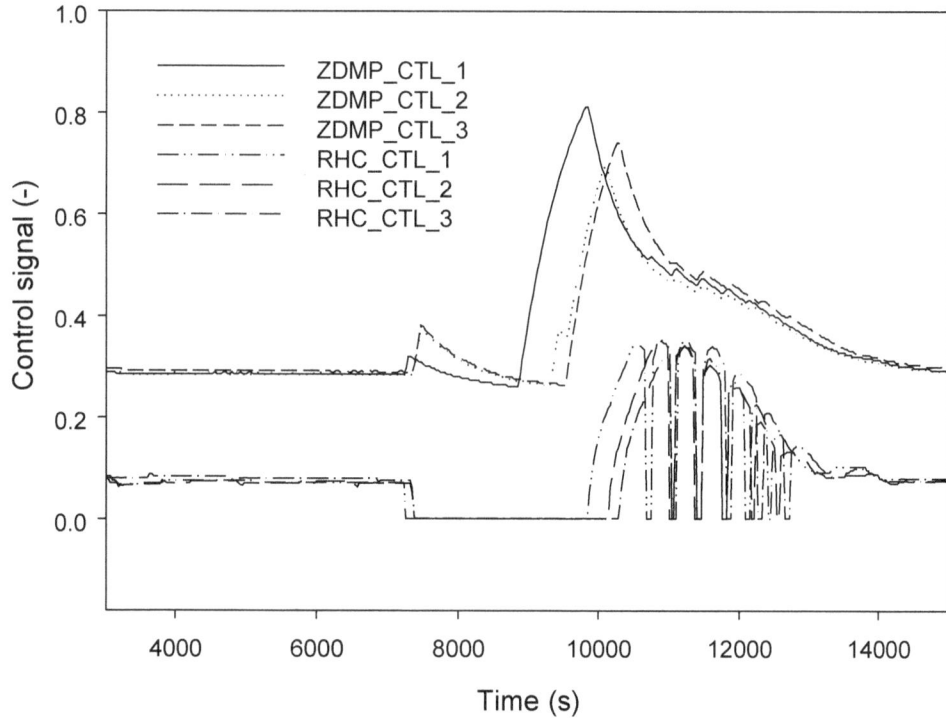

Figure 25. Control signals of AHU1 in Case 1_2.

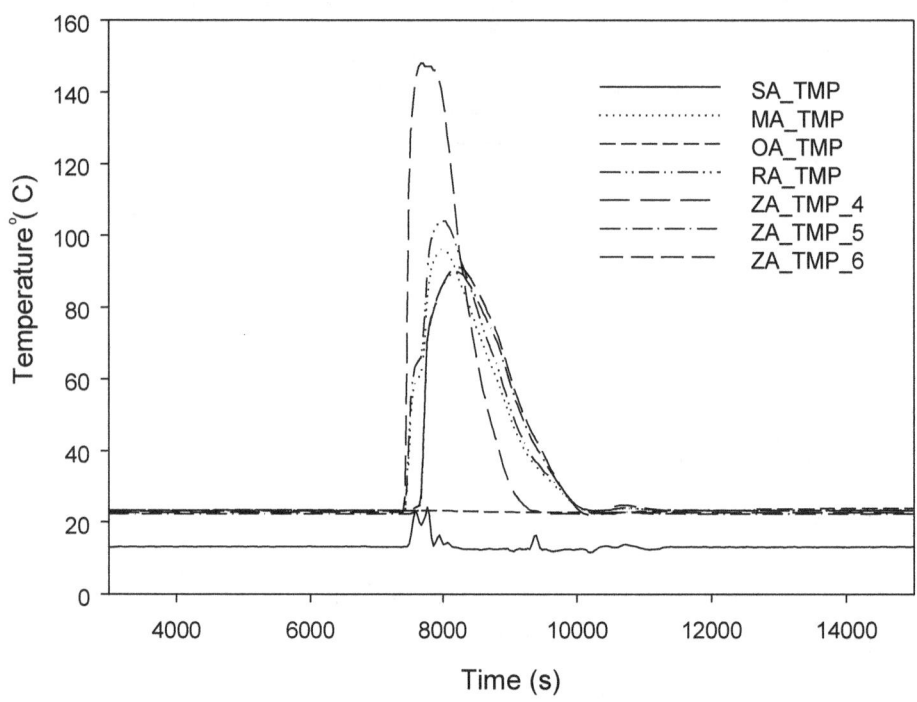

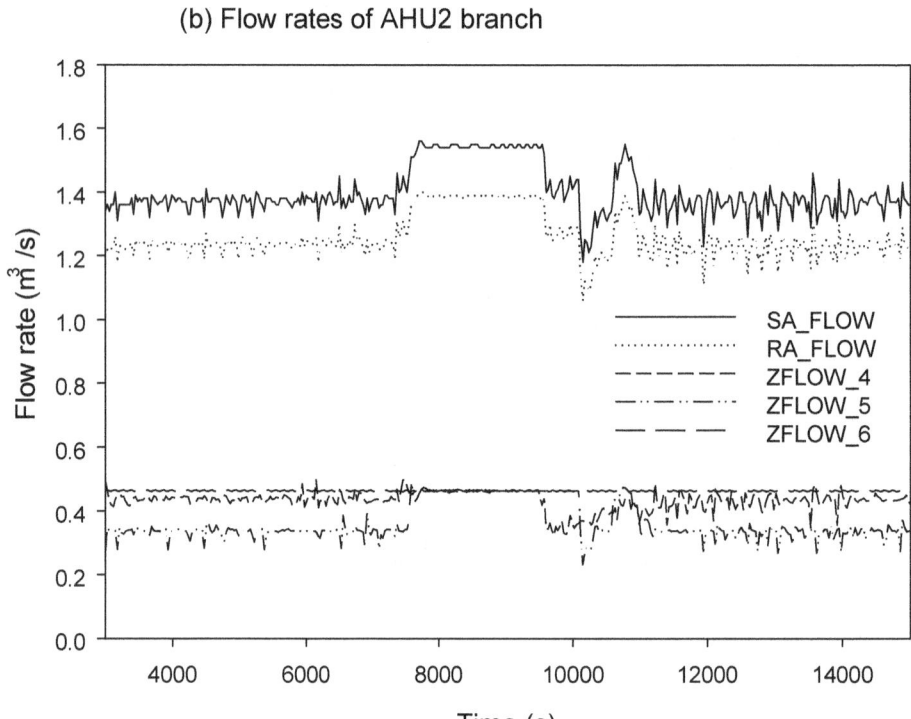

Figure 26. Temperatures, and flow rates of AHU2 in Case 1_2.

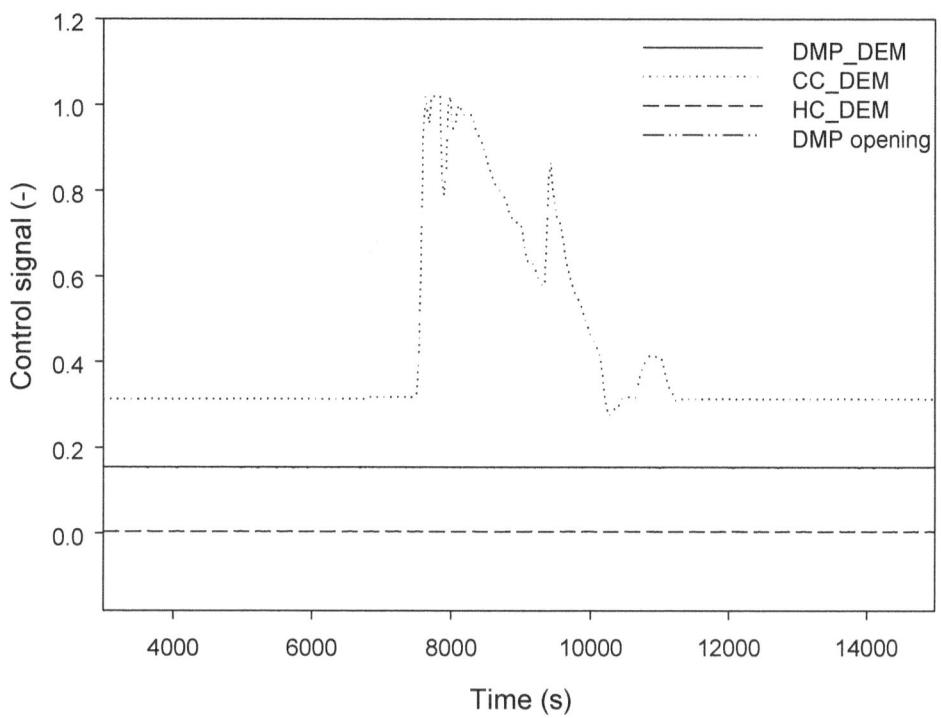

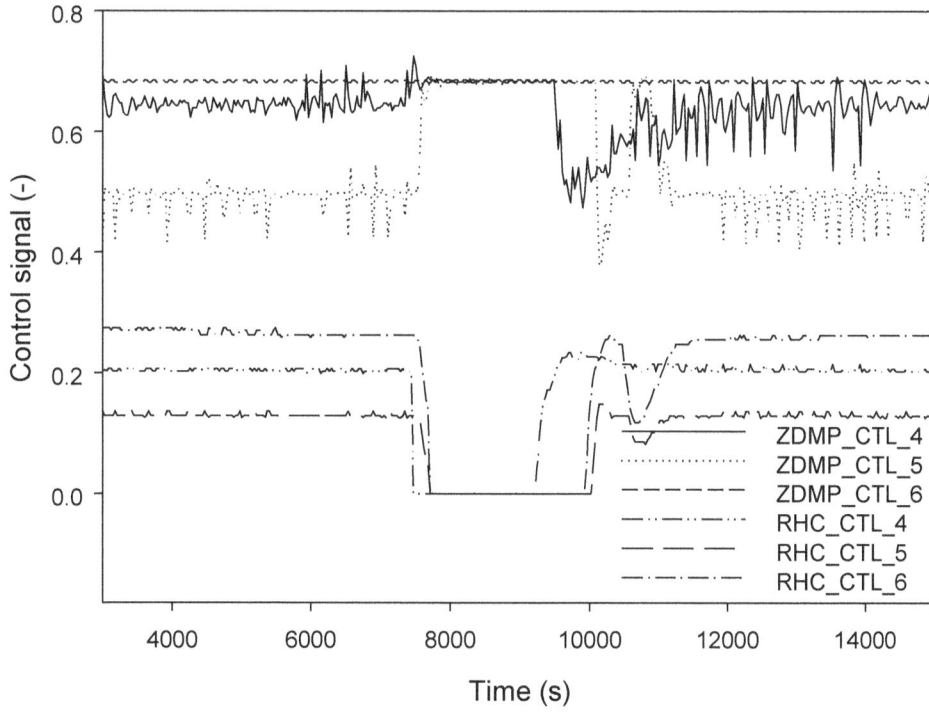

Figure 27. Control signals of AHU2 in Case 1_2.

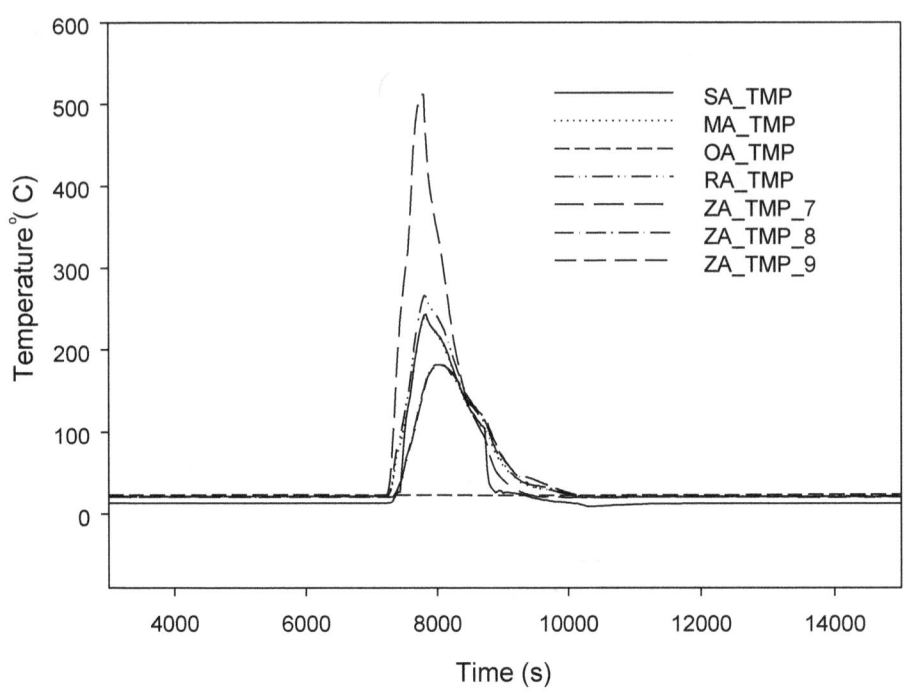

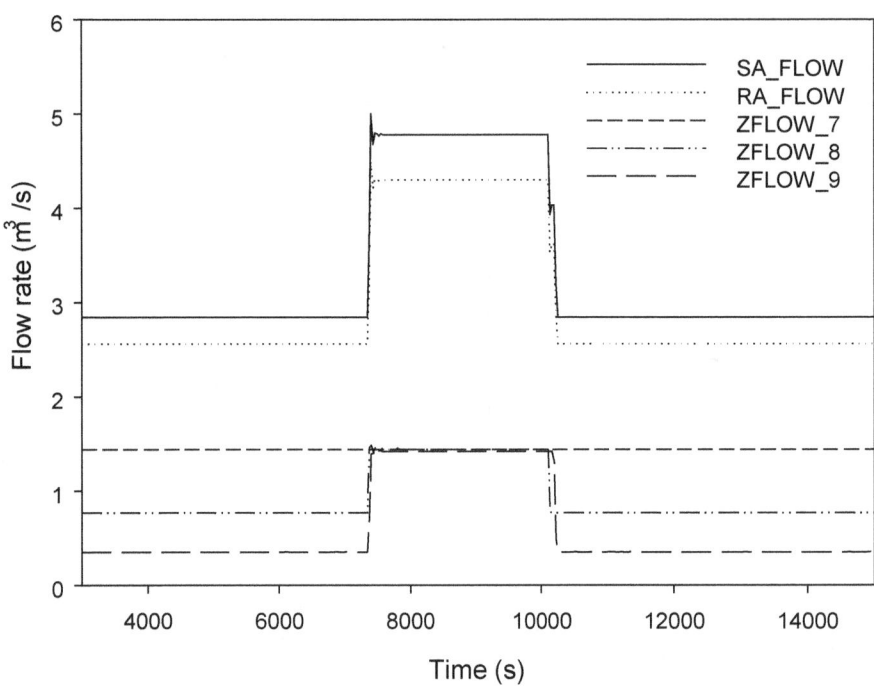

Figure 28. Temperatures, and flow rates of AHU3 in Case 1_2.

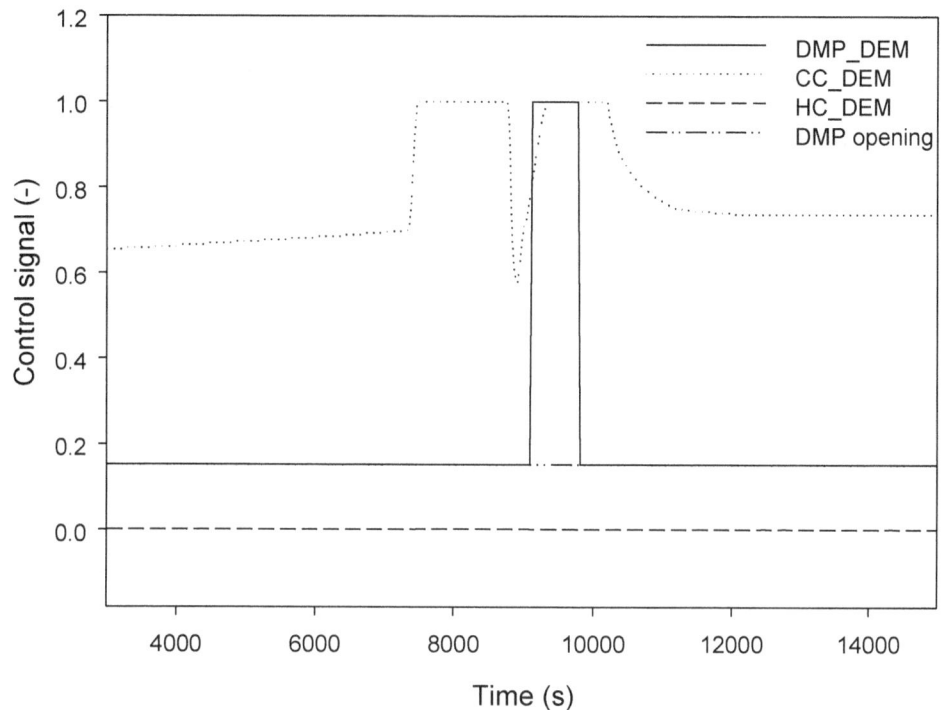

(a) Damper and coil opening demands by AHU3 controller

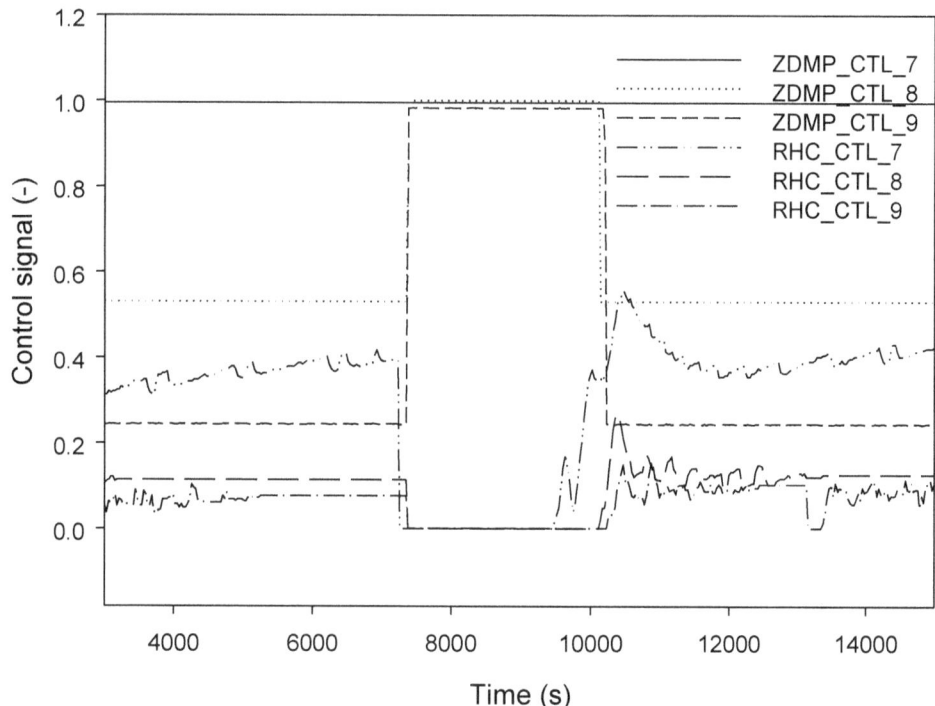

(b) VAV damper and reheat coil signals by VAV controllers

Figure 29. Control signals of AHU3 in Case 1_2.

The fire model, ZFM, was developed based on the modeling approach used in CFAST [25], which models the fire gases as homogeneous upper (hot) and lower (cool) layers. The temperatures and smoke concentrations (or mass fraction) of upper and lower layers are shown in Figure 30 for Case 1_2. The labels Zone_1L, Zone_2L, and Zone_3L represent the lower layer temperatures of Zone 1, 2, and 3, respectively. Likewise Zone_1U, Zone_2U, and Zone_3U represent the upper layer temperatures. As expected, the upper layer temperature of Zone 1 was highest about 1580 °C, and the lower layer temperature of the same zone was 690 °C at its peak. The peak upper and lower layer temperatures of the other two zones were almost 160 °C. The peak values of other two zone's layer temperatures appeared in delay compared with the peak time in Zone 1. The smoke concentrations of Zone 1 apparently were high, while those in other zones, which were located in the second and third floor, were insignificant in this virtual three-story building.

Based upon Bouguer's law, the transmitted light intensity, I, through the smoke is given by:

$$I = I_o e^{-KcL}$$

where

I_o = incident light intensity
K = extinction coefficient
c = smoke density
L = path length

Thus, the path length is expressed as

$$L = \frac{\ln(\frac{I}{I_o})}{-Kc}$$

According to Mulholland and Johnson [20], the extinction coefficient for over-ventilated flaming combustion is 8.7 m^2/g. Using the value of air density, 1294.3 g/m^3, the visibility limit or path length in meters that we see only 0.1 % of light is

$$L = \frac{\ln(0.001)}{-8.7(1294.3 st)}$$

where st is the dimensionless smoke concentration.

Figure 31 shows the visibility limits at the upper layers of the nine conditioned zones of the building and the unconditioned stairwell in Case 1_2. As seen in Figure 31, the visibility was low in zones at the 1st floor (Zone 1, 4, and 7) and the stairwell. U_1 in the figure represents the visibility limit of the upper layer in Zone 1.

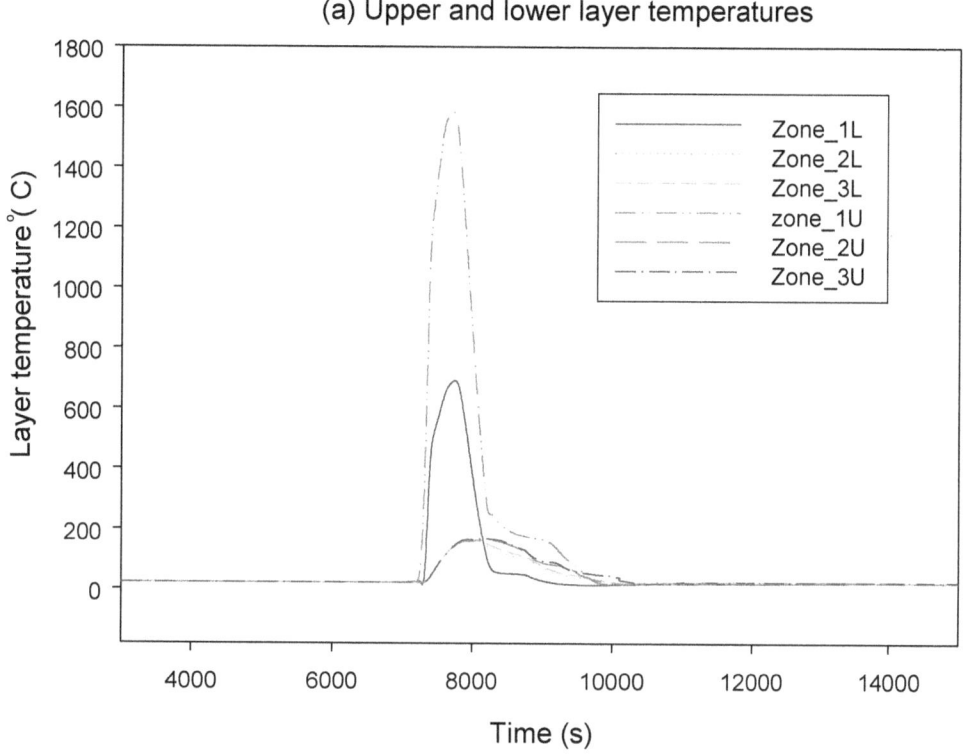

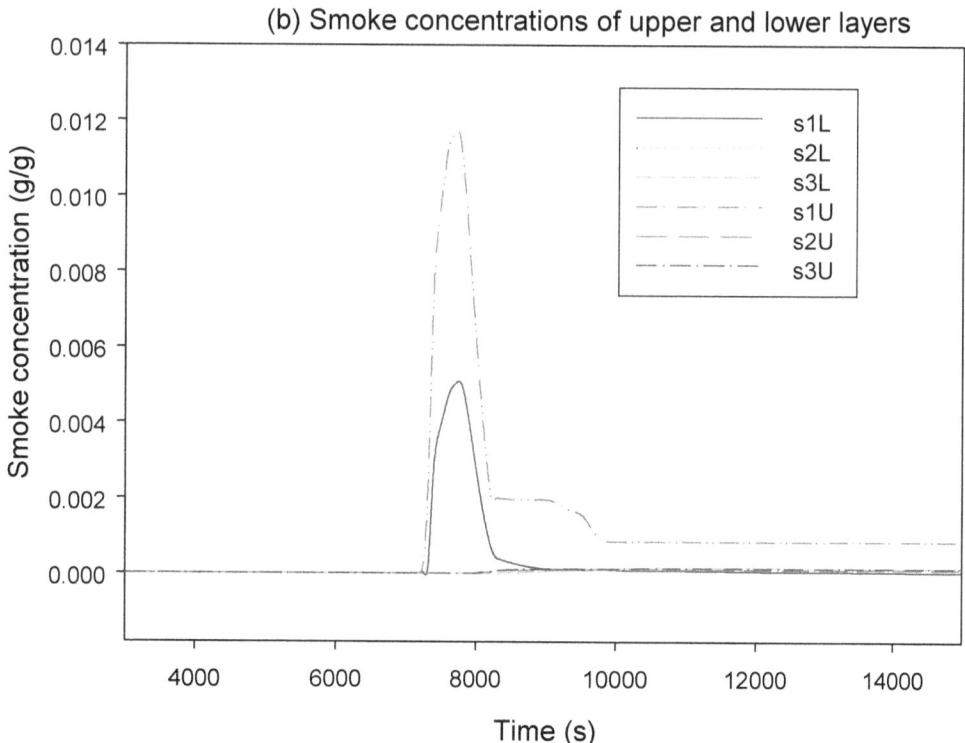

Figure 30. Layer temperatures and smoke concentrations of zones served by AHU1 in Case 1_2.

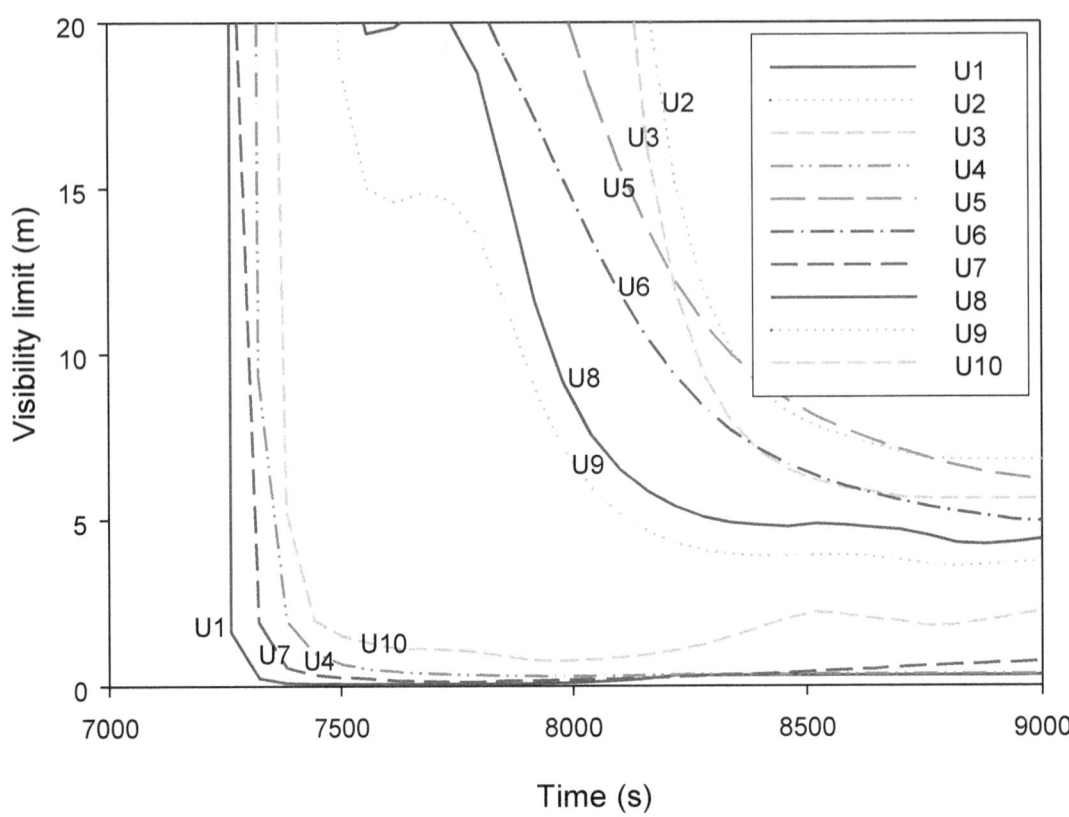

Figure 31. Visibility limits at the upper layers due to smoke in Case 1_2.

An emulation run (Case 1_3) was performed with the same configuration as Case 1_2 except that the AHU damper opening was allowed to obey the BMS enthalpy economizer command. Namely the restriction given to the damper opening to be 15 % at all times was removed. As a result, the AHU damper was opened fully whenever the return air enthalpy was greater than the outside air enthalpy. The temperatures, flow rates, and BMS control signals of AHU1 are shown in Figures 32 and 33. When ZA_TMP_2 and ZA_TMP_3 of Case 1_3 are compared with those of Case 1_2, quite different outcomes can be seen.

The zone air temperatures in Case 1_3 other than in Zone 1 did not significantly increase. Since the most of the return air was exhausted and almost 100 % of fresh air entered into the mixing box when the AHU damper was fully open, the supply air temperature, SA_TMP, and the mixed air temperature, MA_TMP, were considerably lower despite the high temperature of the return air due to fire. Layer temperatures and smoke concentrations of the zones served by AHU1 in Case 1_3 are shown in Figure 34. Compared with Figure 30a, it can be seen that the upper and lower temperatures of Zone 3 are much lower in this case. Those of Zone 1 also are slightly lower. Otherwise Figure 34 is very similar to Figure 30.

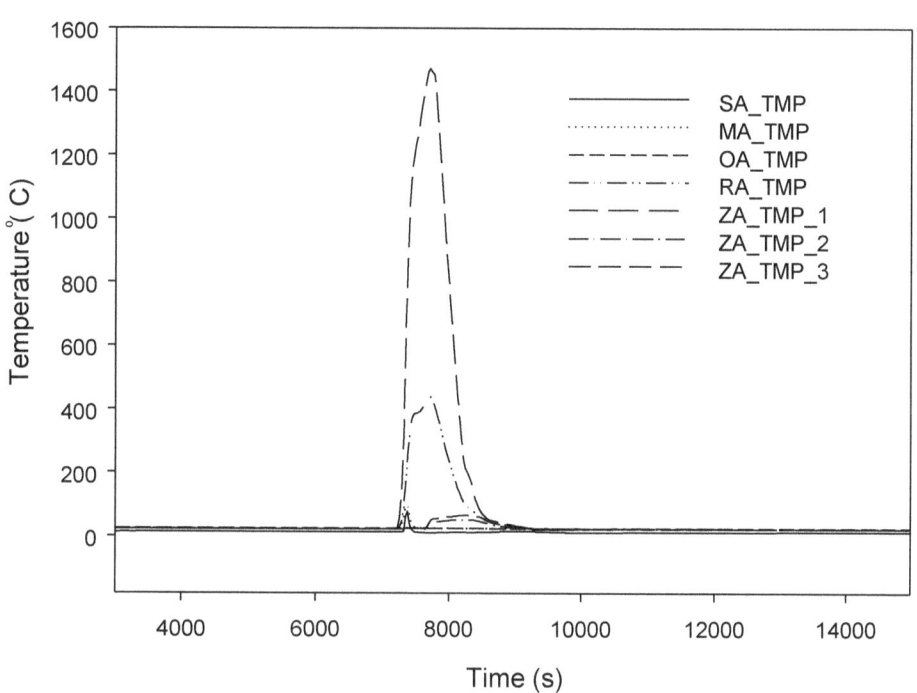

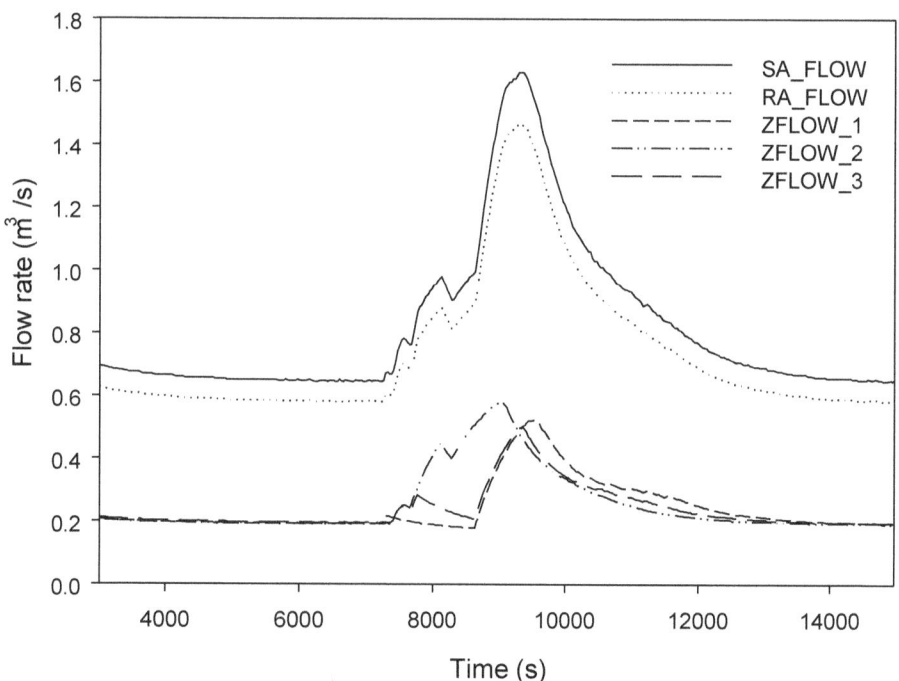

Figure 32. Temperatures, and flow rates of AHU1 in Case 1_3.

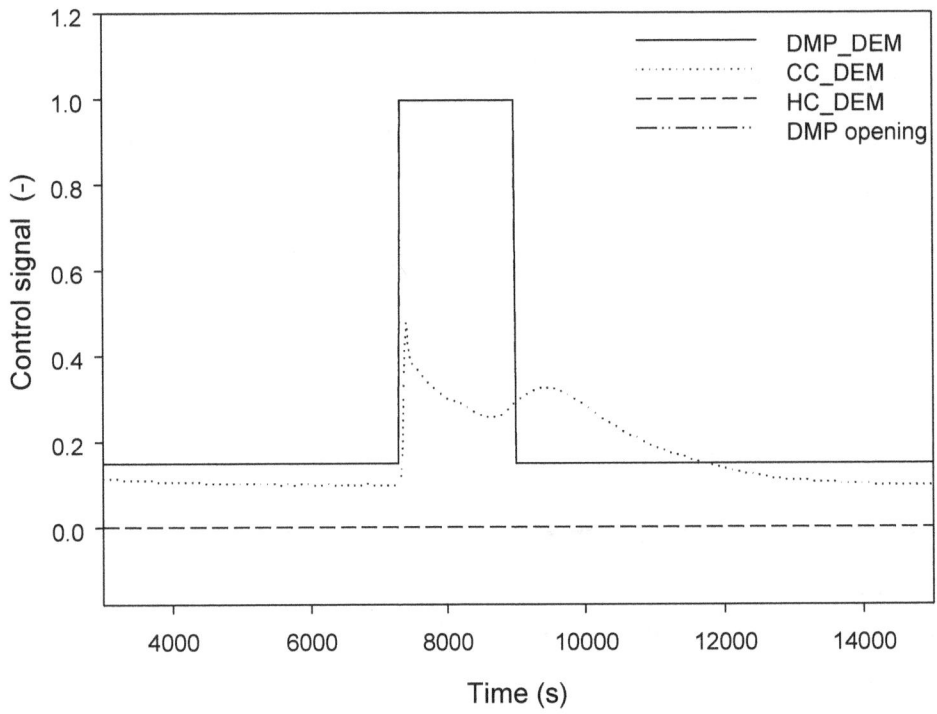

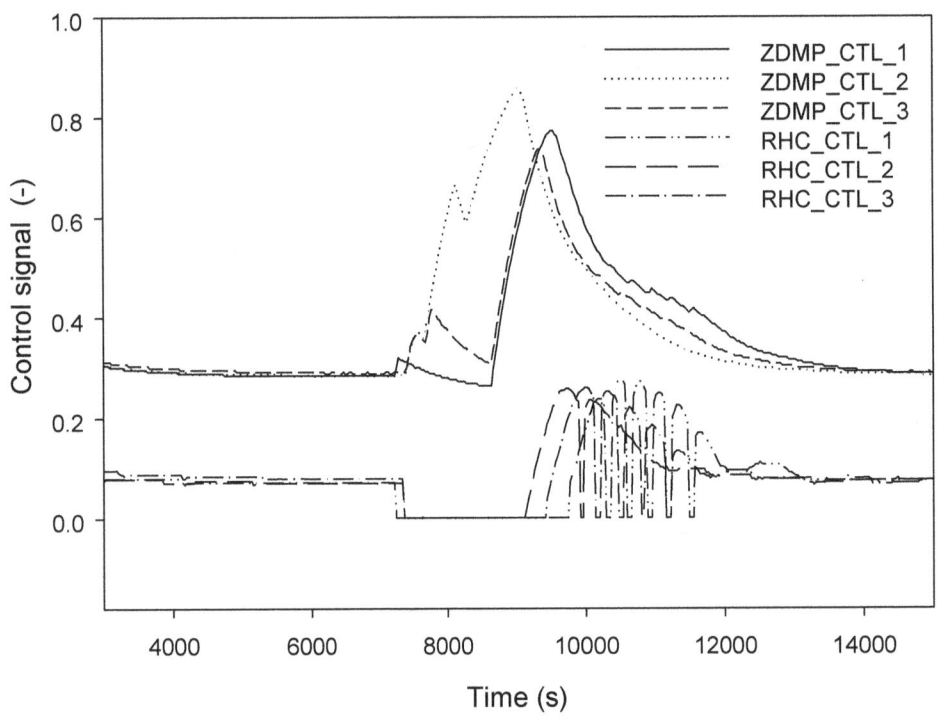

Figure 33. Control signals of AHU1 in Case 1_3.

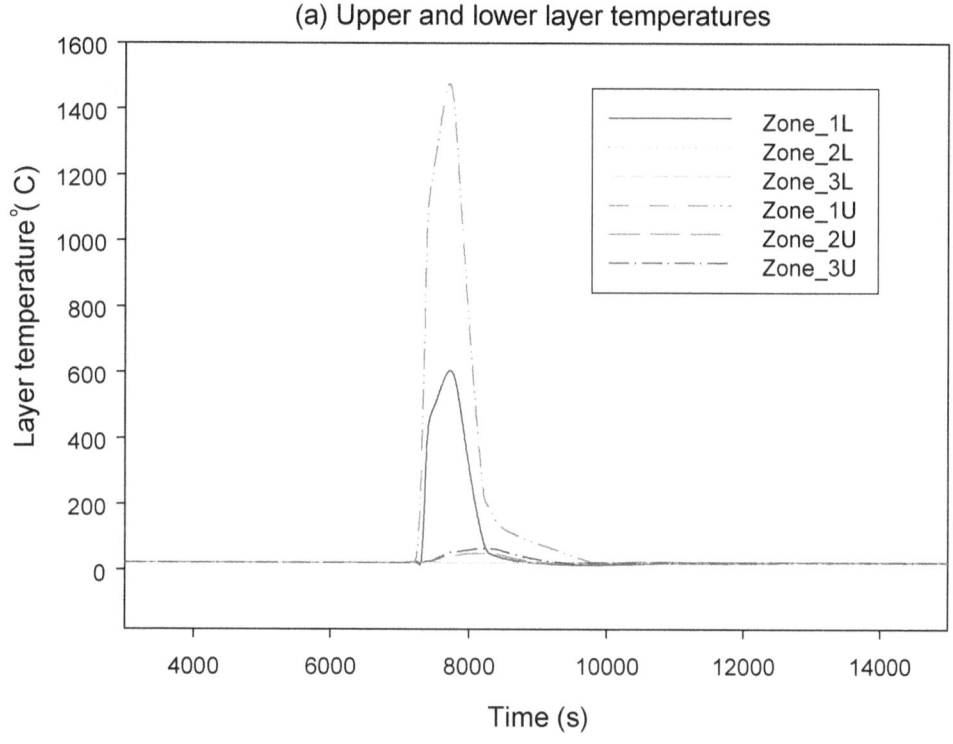

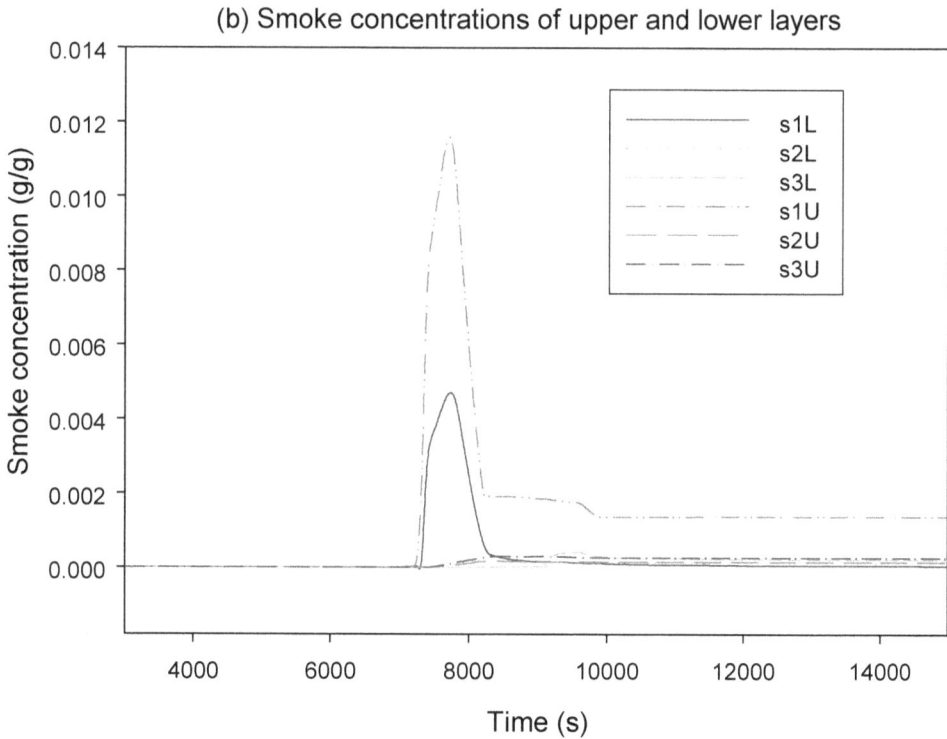

Figure 34. Layer temperatures and smoke concentrations of zones served by AHU1 in Case 1_3.

4.2 Case 2 Scenario

Figure 35 shows a layout of a single-story virtual building. There are eight identically sized rooms (or zones) and a corridor. Each zone has a VAV box and each AHU serves 3 VAV boxes. AHU 1 serves for Zones 1, 2 and 3; AHU 2 for Zones 4, 5 and 6; and AHU 3 for Zones 7, 8 and 9, respectively as seen in Figure 36.

A fire was set in Zone 1 at 7200 s of emulation time. The enthalpy economizer cycle signal from BACnet AHU controller was ignored and set to the minimum damper position. Since all the zones are on the same floor, the fire-spreading pattern appeared quite different from the Case 1. Figures 37 and 38 shows temperatures, airflow rates, and BMS control signals of AHU1. SA_TMP reached up to 960 °C, so did MA_TMP. The return air temperature, RA_TMP, went up to 1070 °C.

Layer temperatures and smoke concentrations of zones served by AHU1 are shown in Figure 39. Obviously the configuration makes quite different form of distribution of temperature as well as smoke compared with the previous scenario. As seen in Figure 40, visibility limits were very low in all the zones. No data of AHU2 and AHU3 are presented here.

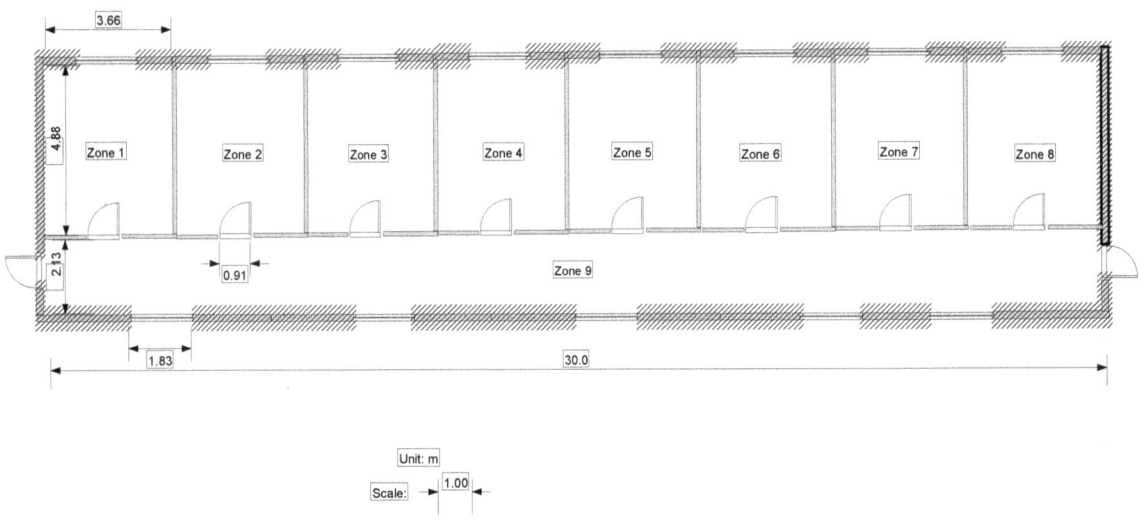

Figure 35. Floor layout of the virtual single-story building with eight identically sized zones in Case 2.

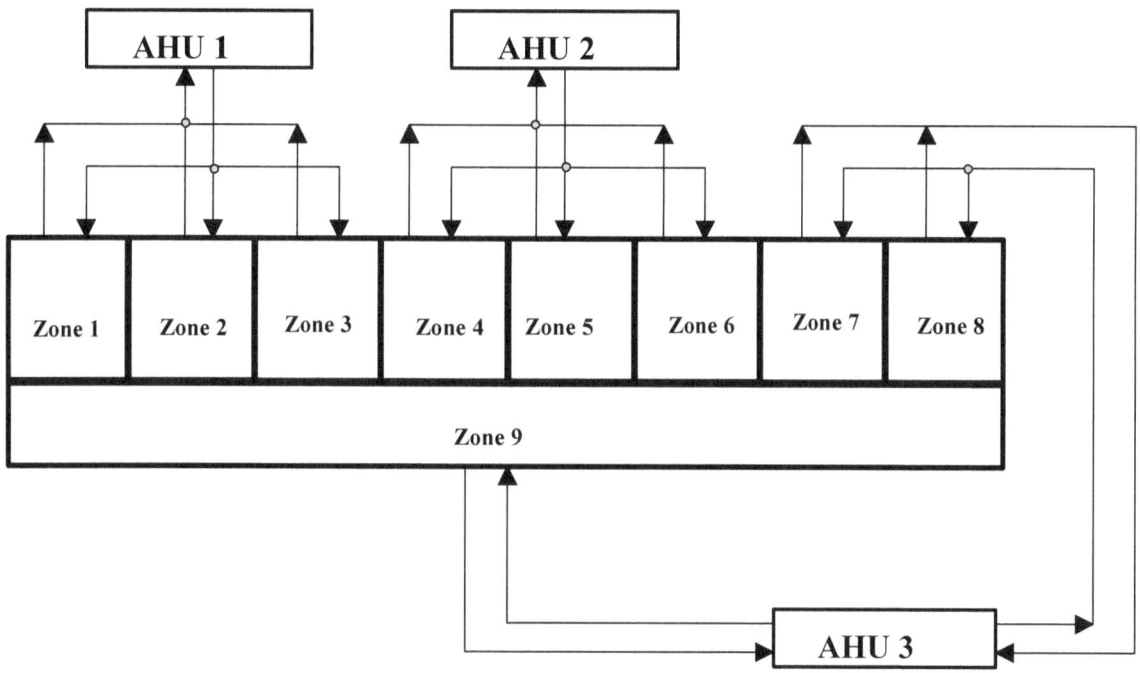

Figure 36. Air distribution for the single-floor building with eight same-size zones in Case 2.

(a) Temperatures at AHU1 and return air temperatures from zones 1, 2, and 3

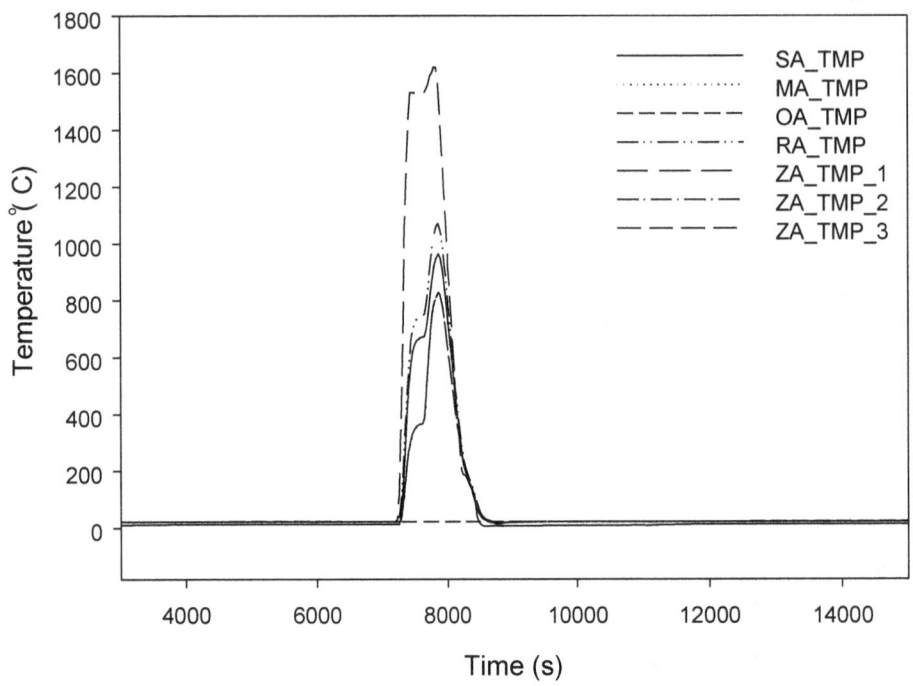

(b) Flow rates of AHU1 branch

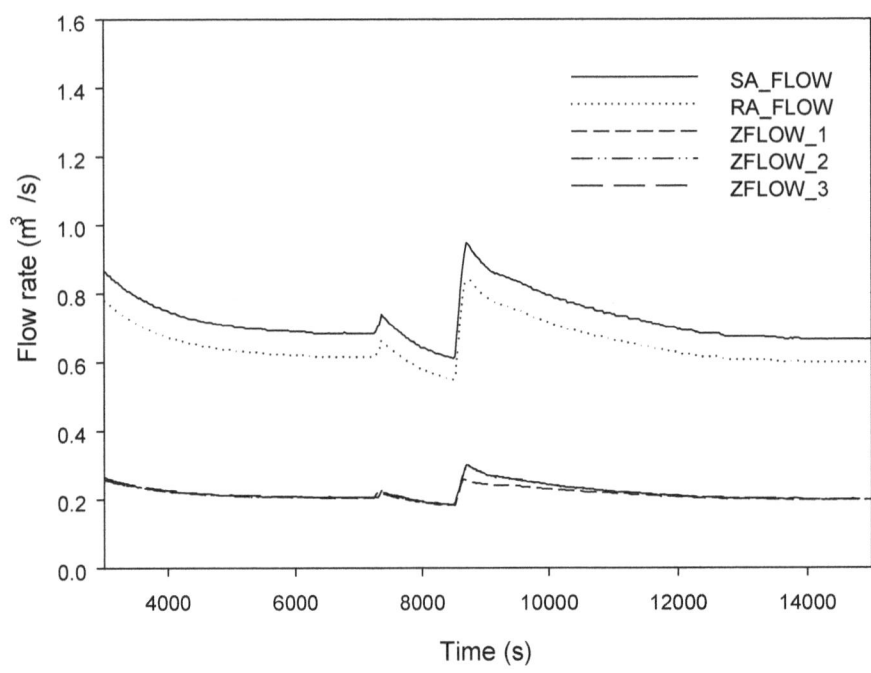

Figure 37. Temperatures, and flow rates of AHU1 in Case 2.

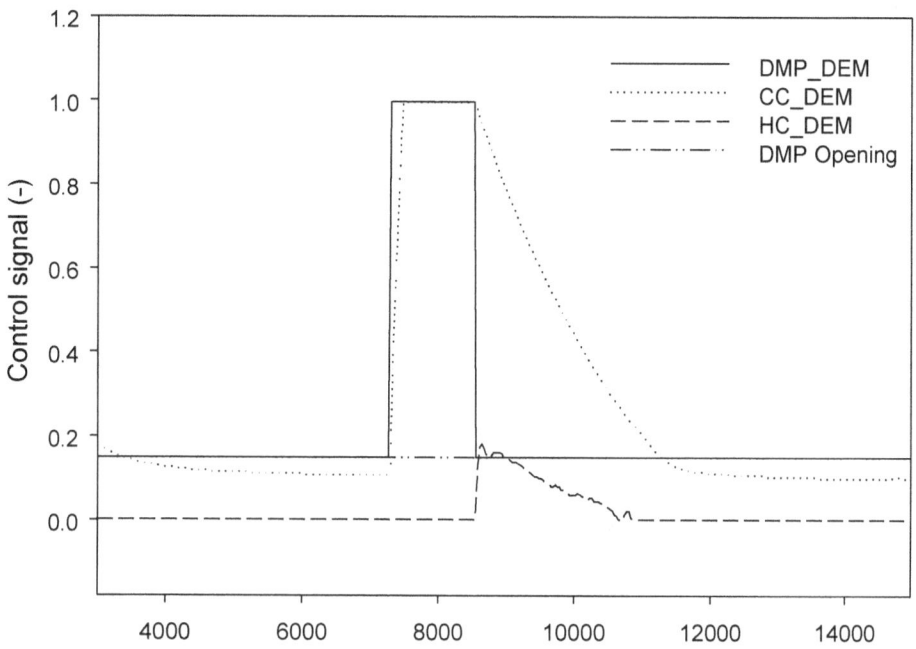

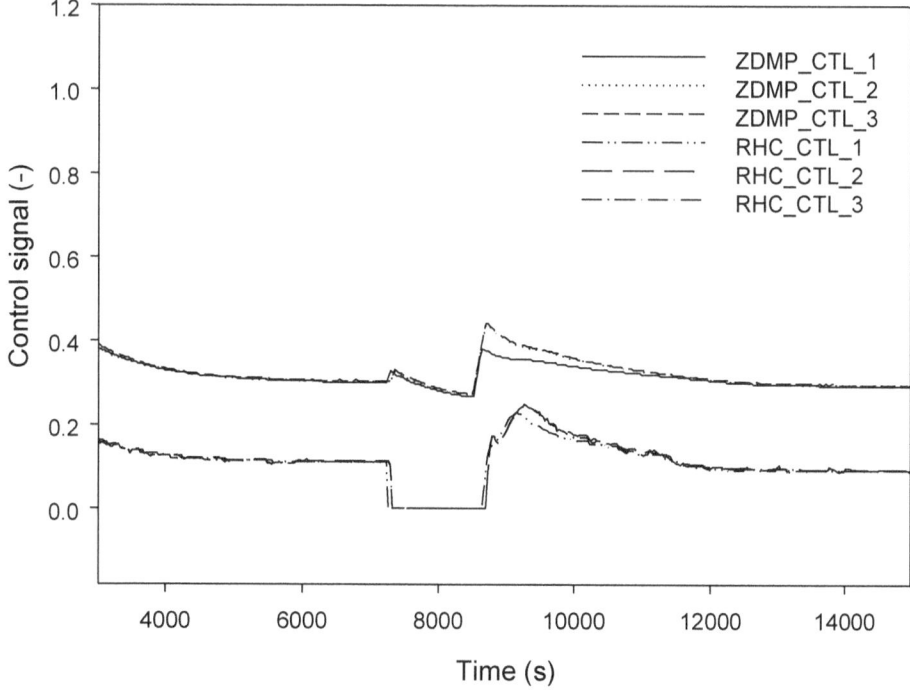

Figure 38. Control signals of AHU1 in Case 2.

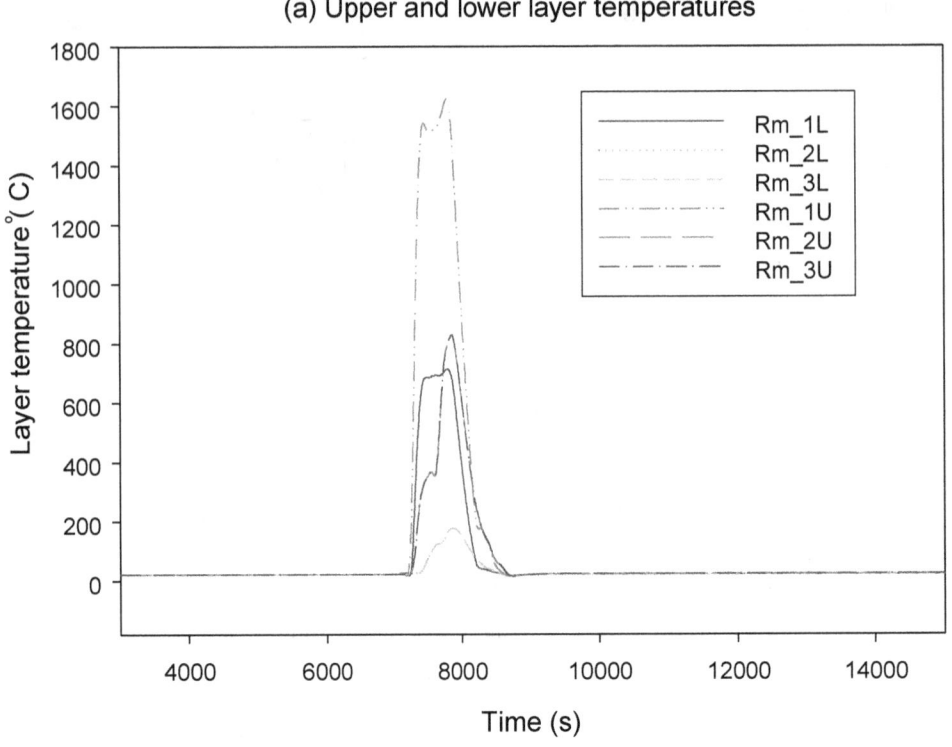

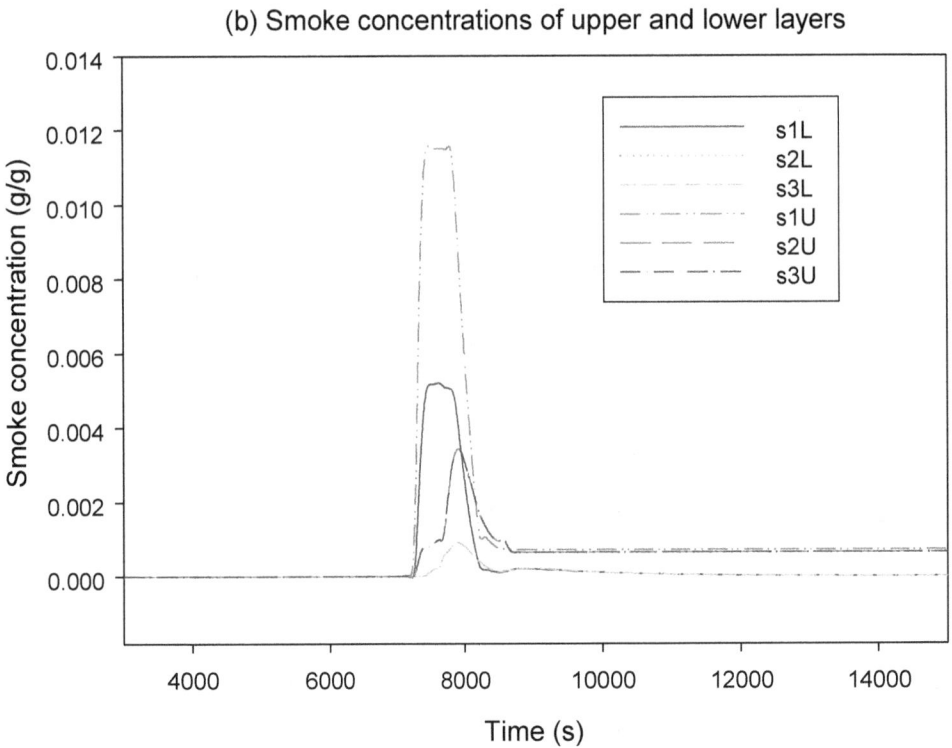

Figure 39. Layer temperatures and smoke concentrations of zones served by AHU1 in Case 2.

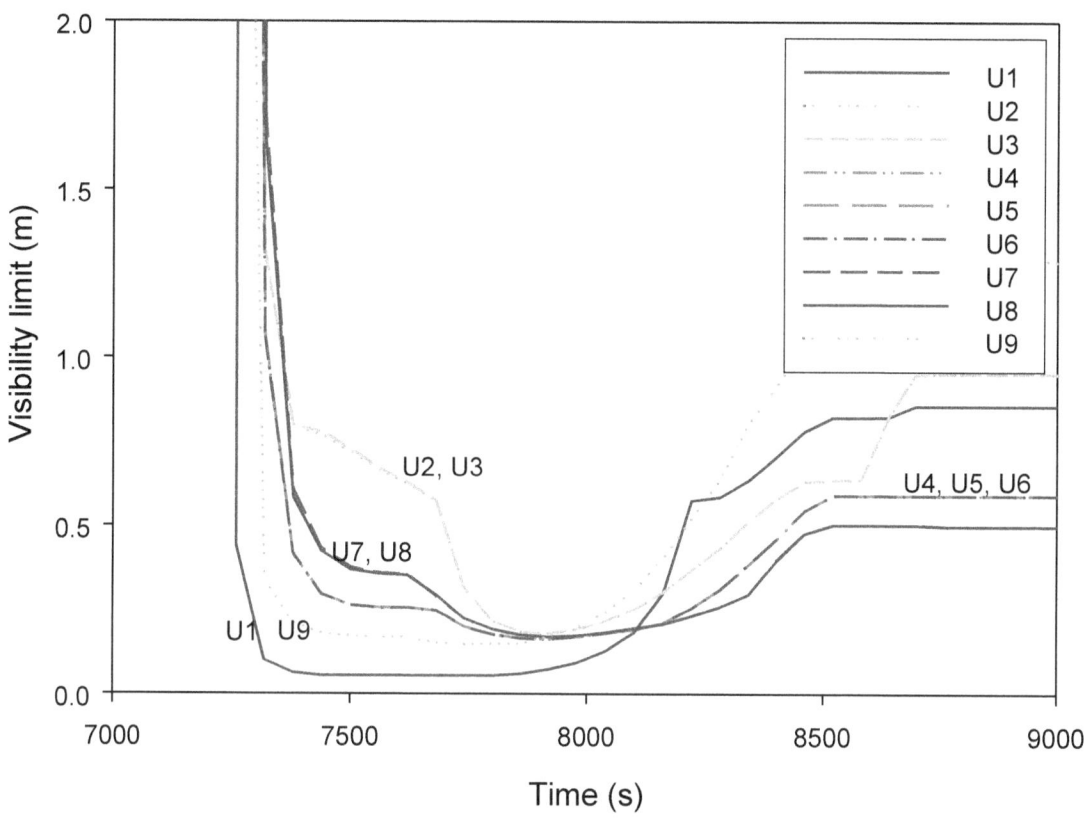

Figure 40. Visibility limits at the upper layers due to smoke in Case 2.

4.3 Case 3 Scenario

A virtual single-story building with nine zones with different sizes was considered. Figure 41 shows the floor layout and Figure 42 shows the air distribution by three air handling units. Each zone has a VAV box and each AHU serves three VAV boxes. AHU1 serves for Zone 1, 2, and 3; AHU2 for Zone 4, 5, and 6; AHU3 for Zone 7, 8, 9, respectively. Zone 8 is a corridor. As was done in Cases 1 and 2, fire was set in Zone 1, and the start time of fire was 7200 s of emulation time. The behavior of AHU1 is close to that in Case 2 as seen in Figures 43 and 44. Due to the different configuration, the time history of upper and lower layer temperature and smoke concentrations as shown in Figure 45 has some differences as expected. Quite low visibility in all zones can be seen in Figure 46.

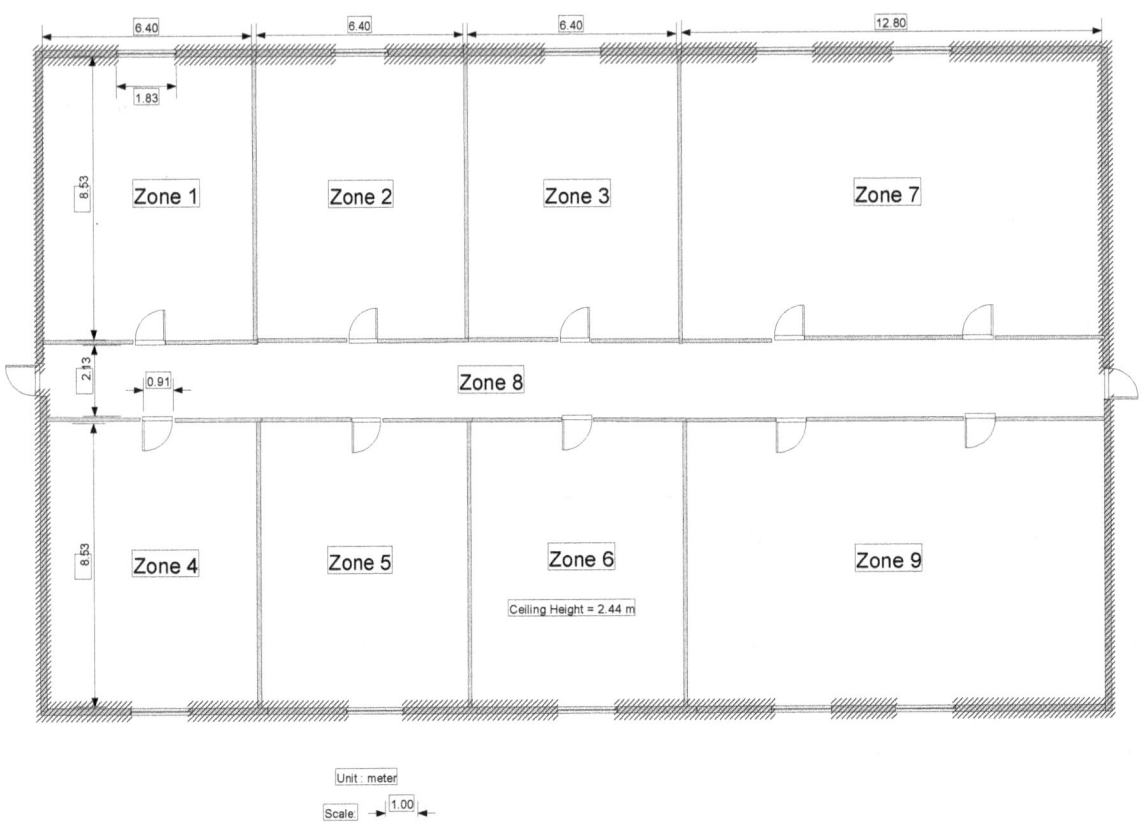

Figure 41. Floor layout of the single-story building with different size zones in Case 3.

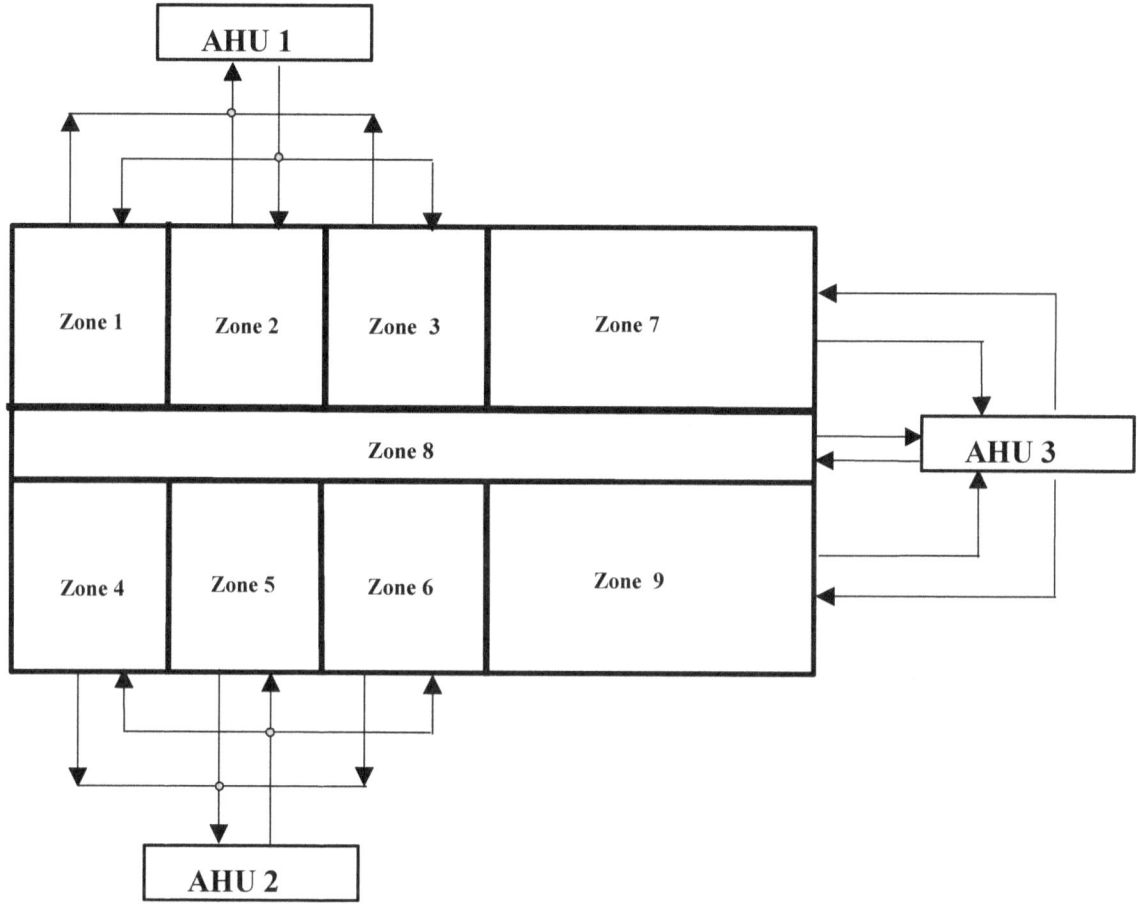

Figure 42. Air distribution for the virtual single-floor building with different size zones in Case 3.

(a) Temperatures at AHU1 and return air temperatures from zones 1, 2, and 3

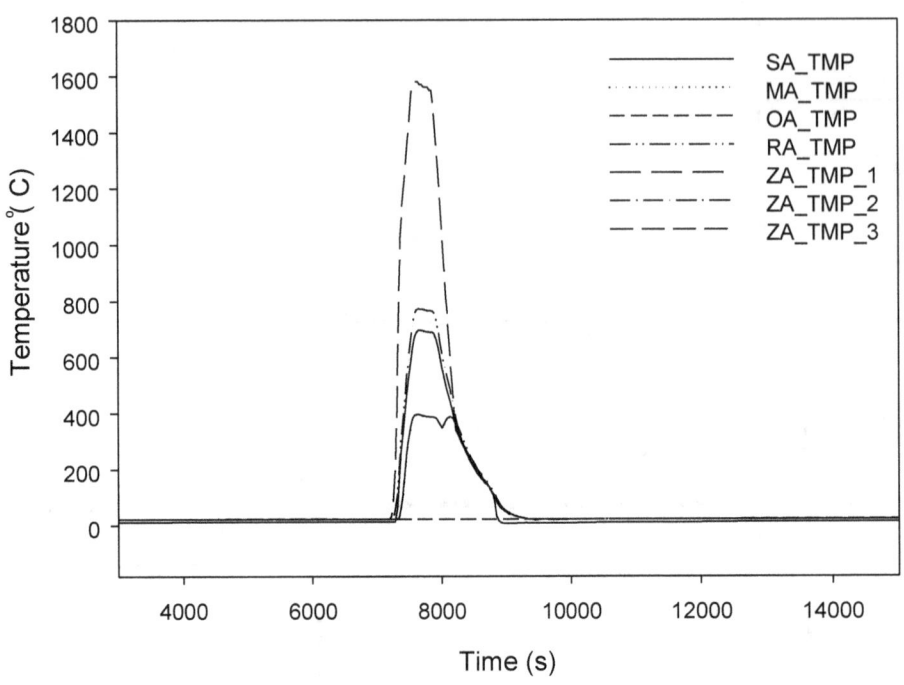

(b) Flow rates of AHU1 branch

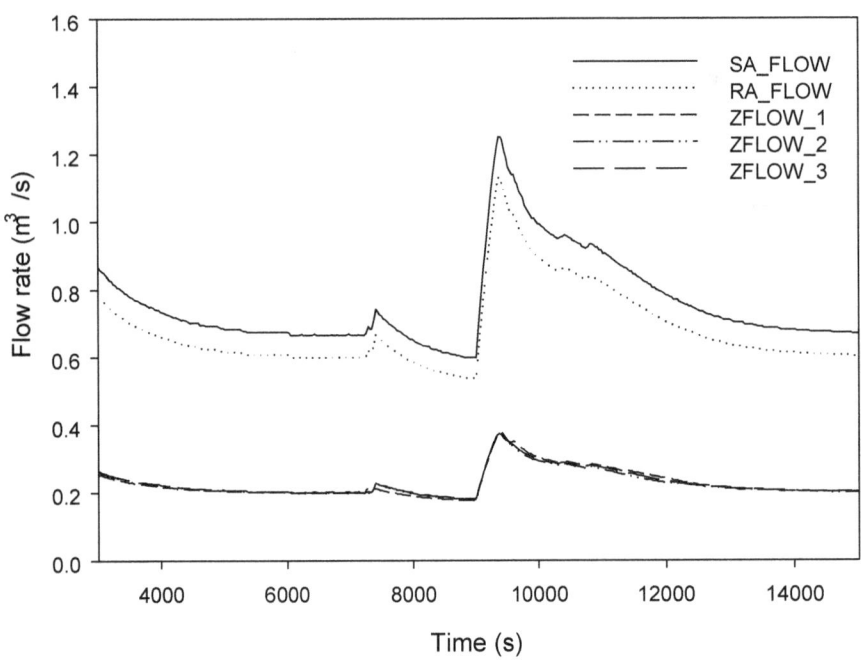

Figure 43. Temperatures, and flow rates of AHU1 in Case 3.

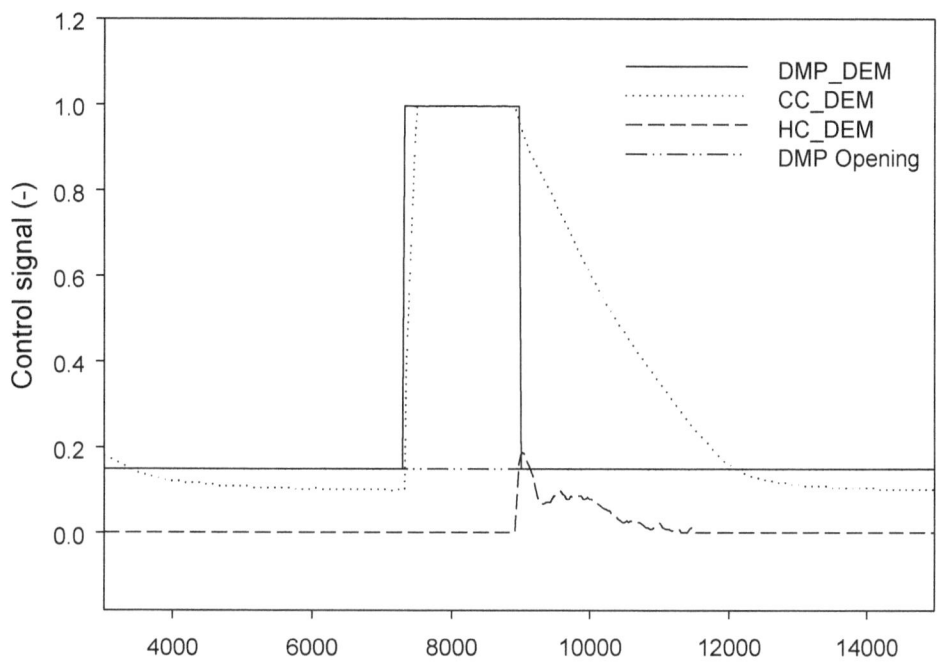

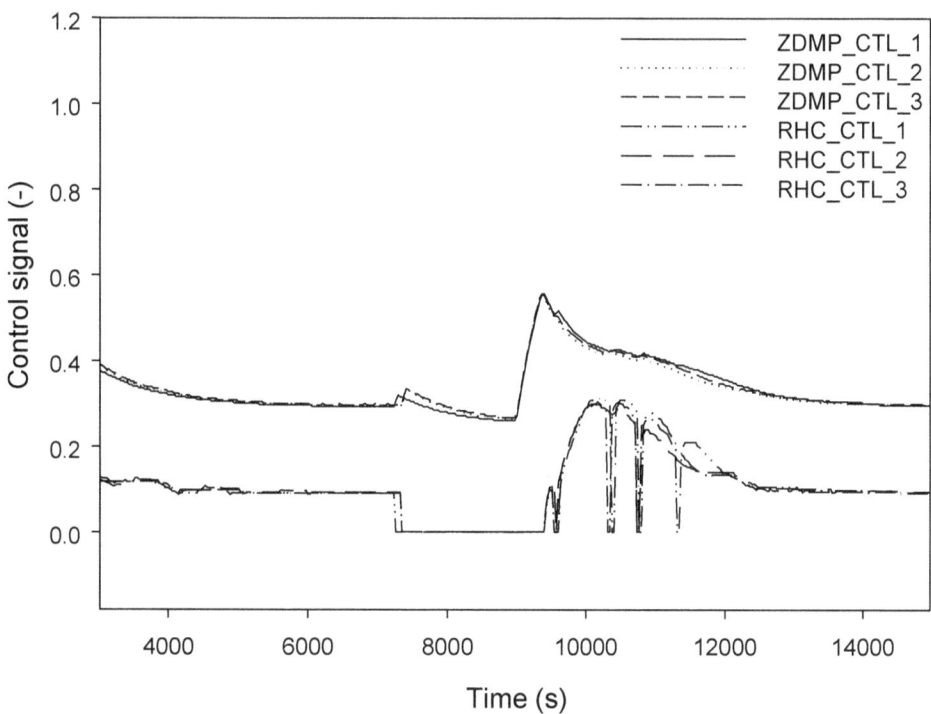

Figure 44. Control signals of AHU1 in Case 3.

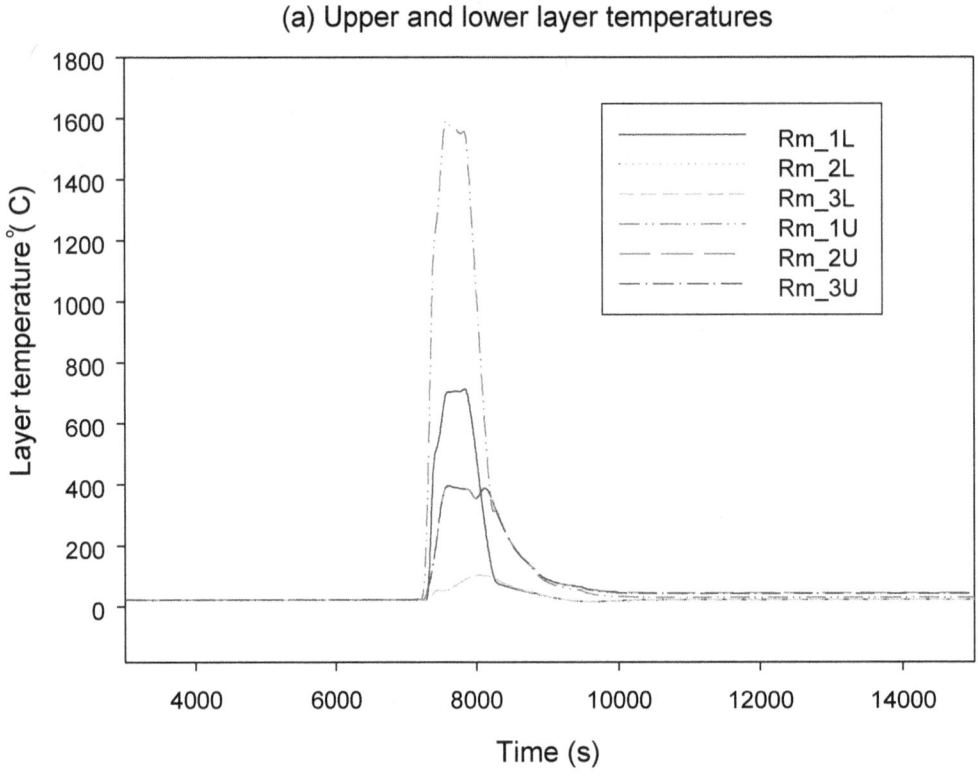

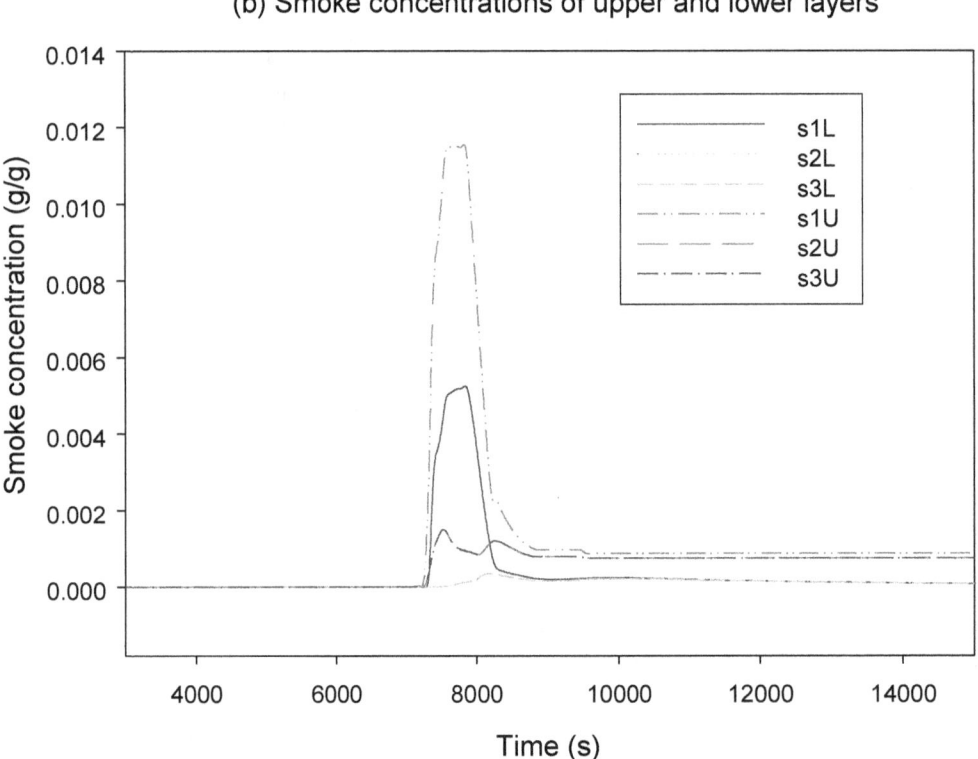

Figure 45. Layer temperatures and smoke concentrations in Case 3.

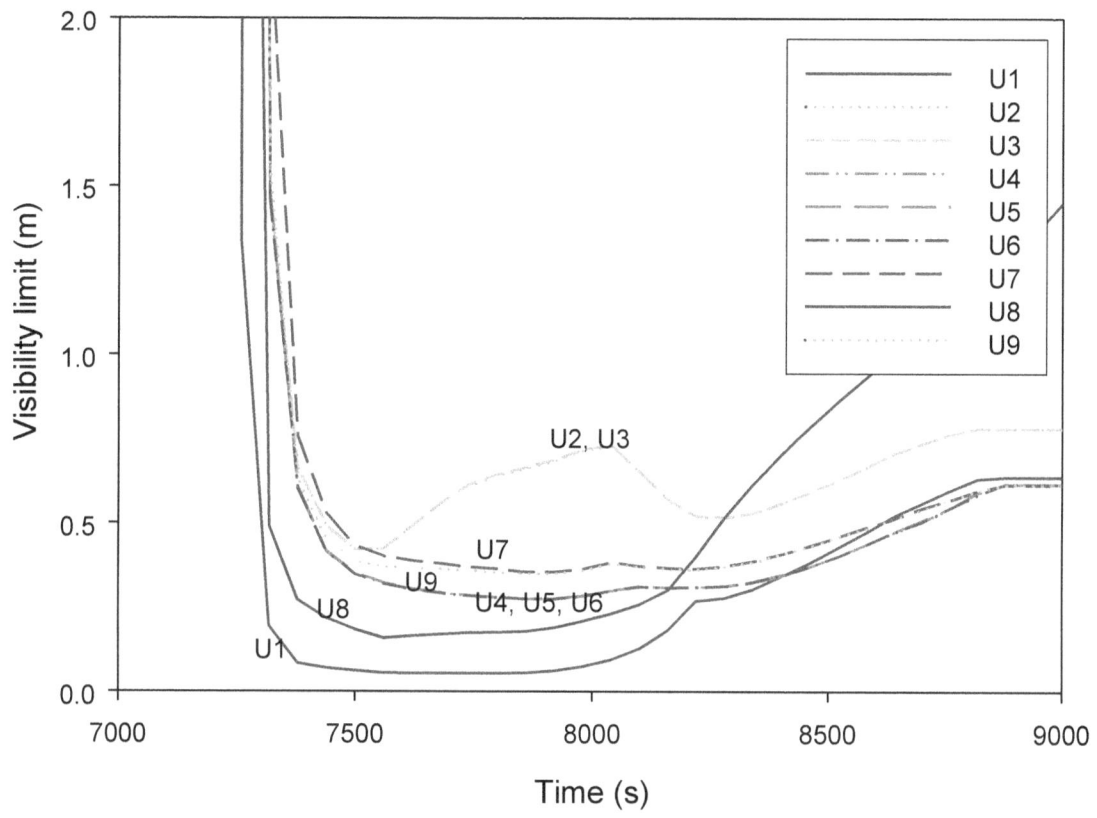

Figure 46. Visibility limits at the upper layers due to smoke in Case 3.

5. CONCLUSION

The VCBT has been enhanced and the overall structure of the current VCBT is reviewed in this preliminary report. ZFM-HVAC was added to the VCBT, the component models used in the HVAC-side of ZFM-HVAC were described and three scenarios were emulated. Some of the emulation results are presented graphically along with brief discussion. Interaction between building HVAC system and fire was demonstrated. In this study, nine conditioned space cases were utilized, because of availability of real BACnet controllers.

Recently, capability to dynamically configure the capacities of AHUs and VAVs based on the zone floor areas was added to ZFM-HVAC. Also simplification of input data to ZFM-HVAC was made. Additionally virtual AHU and VAV box controllers were added to the VCBT in order to emulate buildings with a large number of zones, not limited to nine zones as done this report.

6. REFERENCES

1. May, W.B. & C. Park, Building Emulation Computer Program for Testing of Energy Management and Control System Algorithms, NBSIR 85-3291, NIST, 1985.

2. Kelly, G.E., C. Park, and J.P. Barnett, Using Emulators/Tester for Commissioning EMCS Software, Operator Training, Algorithm Development, and Tuning Local Control Loops, ASHRAE Trans., V. 97, Pt. 1, 1991.

3. Kohonen, R., et al., Development of Emulation Methods, IEA Annex 17 Report, August 1993.

4. Benninn, H., BACnet Virtual Building, Practical Training Report, University of Maryland, 1997.

5. Decious, G.M., C. Park, & G.E. Kelly, A Low-Cost Building/HVAC Emulator, Heating/Piping/AirConditioning, HPAC, Jan. 1997.

6. Fuhrmann, F., Integrating Multiple Programs over the Internet using CORBA, Practical Training Report, University of Maryland, 1999.

7. Bushby, S.T., N.S. Castro, M. Galler, C. Park, and J.M. House, Using the Virtual Cybernetic Building Testbed and FDD Test Shell for FDD Tool Development, NISTIR 6818, NIST, 2002.

8. Park, C., D.R. Clark, and G.E. Kelly, An Overview of HVACSIM$^+$, A Dynamic Building/HVAC/Control Systems Simulation Program, Proceedings of the 1st Annual Building Energy Simulation Conf., Seattle, WA, August 21-22, 1985.

9. Castro, N.S., J. Schein, C. Park, M.A. Galler, S.T. Bushby, & J.M. House, Results from Simulation and Laboratory Testing of Air Handling Unit and variable Air Volume Box Diagnostic Tools, NISTIR 6964, NIST, 2003.

10. Bushby, S.T., Application Layer Communication Protocols for Building Energy management and Control Systems, ASHRAE Transactions 1988, V. 94, Pt. 2, ASHRAE, 1988.

11. Bushby, S.T., BACnet: A Standard communication Infrastructure for Intelligent Buildings, Automation in Construction, V. 6, No. 5-6, 1997.

12. Reneke, P.A., R.D. Peacock, G.P. Forney, and W.D. Davis, ZFM Zone Fire Model (Version 1) Reference Guide, (to be published in NIST publication).

13. Siegel, J., CORBA Fundamentals and Programming, John Wiley & Sons, 1996.

14. Norford, L.K. and P. Haves, A Standard Simulation Testbed for the Evaluation of Control algorithms and Strategies, ASHRAE Report 825-RP, ASHRAE, 1997.

15. Kusuda, T., Thermal Response Factors for Multilayer Structures of Various Heat Conduction Systems, ASHRAE Trans. V. 75, 1969, pp. 246-270.

16. WYEC2 User's Manual and Toolkit, ASHRAE contract 728-TRP, Augustyn+ Company, 1997.

17. Seborg, D.E., T.F. Edgar, and D.A. Mellichamp, Process Dynamic and Control, John Wiley & Sons, 1989.

18. Treado, S., S.C. Cook, & M. Galler, Distributed Biometric Access Control Testbed, NISTIR 7190, NIST, 2004.

19. Peacock, R.D., G.P. Forney, P. Reneke, R. Portier, and W.W. Jones, CFAST, the Consolidated Model of Fire Growth and Smoke Transport, NIST Tech. Note 1299, NIST, 1993.

20. Mulholland, G.W., and E.L. Johnson, A New Smoke Concentration Meter, 3rd International Conf. on Fire Research and Engineering, Chicago, IL, Oct. 4-8, 1999.

Appendix A. Sample BACnet Air-handling Unit Controller Object Database.

BACnet Objects of a VAV Air-handling Unit Controller			
Object	Description	Default	Remarks
AI-0	Supply Air Temperature	n/a	0 to 5 volts = 0 °F to 200 °F
AI-1	Supply Air Pressure	n/a	0 to 5 volts = 0" to 5"
AI-2	Return Air Temperature	n/a	0 to 5 volts = 0 °F to 100 °F
AI-3	Return Air Humidity	n/a	0 to 5 volts = 0 % to 100 %
AI-4	Outside Air Temperature	n/a	0 to 5 volts = -50 °F to 150 °F
AI-5	Outside Air Humidity	n/a	0 to 5 volts = 0 % to 100 %
AI-6	Supply Flow Rate	n/a	0 to 5 volts = 0 to 8000 CFM
AI-7	Return Flow Rate	n/a	0 to 5 volts = 0 to 8000 CFM
AI-8	Mixed Air Temperature	n/a	0 to 5 volts = -50 °F to 150 °F
BI-9	Supply Flow Proof	n/a	0 volts = Fan ON, 5 volts = Fan OFF
BI-10	Return Flow Proof	n/a	0 volts = Fan ON, 5 volts = Fan OFF
BO-0	Supply Fan	n/a	
BO-1	Return Fan	n/a	
AO-0	Supply Fan Speed	0	0 to 10 volts = 0 % to 100 %
AO-1	Cooling Coil	0	0 to 10 volts = 0 % to 100 %
AO-2	Heating Coil	0	0 to 10 volts = 0 % to 100 %
AO-3	Economizer	0	0 to 10 volts = 0 % to 100 %
AO-4	Return Fan Speed	0	0 to 10 volts = 0 % to 100 %
AV-0	Temperture Ctrl Signal	0	
AV-28	OSA Enthalpy	0	
AV-29	Return Enthalpy	0	
AV-40	Min. Return Fan Speed	20	
AV-41	Min. Supply Fan Speed	20	
AV-42	Supply/Return CFM Differential	1000	
AV-43	Dehumidifying Supply Temp. Setpoint	50	
AV-44	Current Supply Temp. Setpoint	0	
AV-45	Supply Temp. SP Reset Loop	0	
AV-46	Highest Cooling Signal (from VAV's)	0	
AV-47	Supply Pressure SP Reset Loop	0	
AV-48	Current Supply Pressure Setpoint	0	
AV-49	Highest Need More Air Signal (from VAV's)	0	
AV-50	Supply Pressure Setpoint - Warmup	1.4	
AV-51	Supply Pressure Setpoint - Hi Limit	2	
AV-52	Supply Pressure Setpoint - Lo Limit	0.2	
AV-53	Supply Pressure Setpoint - Manual/Startup	1	
AV-55	Supply Temp. Setpoint - Hi Limit	65	
AV-56	Supply Temp. Setpoint - Lo Limit	50	
AV-57	Supply Temp. Setpoint - Manual/Startup	60	
AV-58	Supply Temp. Setpoint - Warmup	160	
AV-60	Supply Temp. Low Limit	50	
AV-66	Econ Min. Position %	15	
AV-80	Humidity Hi Limit	60	
BV-20	Alarm Reset	OFF	
BV-26	Supply Fan Alarm	OFF	
BV-28	Return Fan Alarm	OFF	
BV-32	Economizer Lockout Status	OFF	ON = Econ. locked out
BV-40	Occupied Command	OFF	Starts AHU, Econ. to Min. Position
BV-41	WarmupCommand	OFF	Starts AHU, Econ. stays closed
BV-43	Return Fan Proof	OFF	Transferred from BI-9
BV-46	Supply Fan Proof	OFF	Transferred from BI-10
BV-47	Supply Pressure Setpoint Reset Switch	OFF	ON = Autoreset, OFF = Manual
BV-48	Supply Temp. Setpoint Reset Switch	OFF	ON = Autoreset, OFF = Manual
BV-49	Dehumidifying	OFF	Indicates system is dehumidifying

Appendix B. *Sample BACnet VAV Box Controller Object Database.*

BACnet Objects of a VAV Box Controller

Object	Description	Default	Remarks
AI-0		n/a	not used
AI-1	Room Temp.	n/a	0 to 5 volts = 30 $^\circ$F to 130 $^\circ$F
AI-2	Discharge Air Temp.	n/a	0 to 5 volts = 40 $^\circ$F to 190 $^\circ$F
AI-3	Airflow Rate	n/a	0 to 5 volts = 0 to 4000 CFM
AO-0	Damper Command	OFF	0 to 10 volts = Full closed to full open
AO-1	Heating Valve Command	OFF	0 to 10 volts = Full closed to full open
AO-2	OFF		not used
AV-0	Heating Signal	0	
AV-1	Cooling Signal	0	
AV-4	Desired Airflow - Cold	0	CFM
AV-6	Occupied Heating Setpoint	0	
AV-7	Occupied Cooling Setpoint	0	
AV-9	Cooling Damper Position %	0	
AV-10	Need More Air - Clg	0	
AV-14	Heating Valve Position %	0	
AV-20	Windsock - Cold (0 to 1)	0	
AV-64	Cooling Damper Time	60	seconds
AV-65	Heating Valve Time	60	seconds
AV-66	Reheat Air - CFM	300	Desired airflow when htg valve full open
AV-67	Max Airflow - CFM	1000	
AV-68	Min. Airflow - CFM	100	Min. air applies only when occupied
AV-90	Occupied Setpoint	72	Limited by AV-91 and AV-92
AV-91	Occupied Setpoint - Hi Limit	80	Limit for AV-90
AV-92	Occupied Setpoint - Lo Limit	65	Limit for AV-90
AV-93	Cooling Offset	1	
AV-94	Heating Offset	1	
AV-95	Unoccupied Cooling Setpoint	85	
AV-96	Unoccupied Heating Setpoint	55	
AV-97	Afterhours Timer Limit	0	Duration for Microtouch
AV-98	Afterhours Timer (hr)	0	Automatically counts down to 0
AV-99	Current Cooling Setpoint	0	Calculated
AV-100	Current Heating Setpoint	0	Calculated
AV-101	Microset Room Temp.	0	Use DDC to transfer AI to this AV
AV-102	Humidity from Microset	0	
AV-103	Outside Air Temp.	0	Transfer from Lsi
AV-104	Mictouch Lever Offset	0	Calculated from current lever position
AV-105	Mictouch Bias Limit	0	Assigned Offset for full stroke
AV-106	Demand Offset	0	Transfer from Lsi
BV-0	Bad Sensor Alarm	OFF	
BV-1	Heating/Cooling Mode: On= Heating	OFF	
BV-2	Persistent Communication Fail	OFF	
BV-7	Main Air Status - Cold	OFF	
BV-8	Warm Air in Duct	OFF	
BV-9	Force Min. Air - Cold	OFF	
BV-10	Force Reheat Air	OFF	
BV-11	Force Max. Air - Cold	OFF	
BV-12	Force Open - Cold	OFF	
BV-13	Force Closed - Cold	OFF	
BV-19	Day Mode 2 hour Delay	OFF	
BV-24	Space Too Warm	OFF	
BV-25	Space Too Cold	OFF	
BV-38	Heating Lockout	OFF	Used for demand control
BV-39	Cooling Lockout	OFF	Used for demand control
BV-40	Occupied Command	OFF	Typically sent from Lsi
BV-41	Optimum Start Heating	OFF	
BV-42	Optimum Start Cooling	OFF	
BV-64	Occupied SP's Command	OFF	Turns ON Micset and BV-67
BV-65	OFF Mode Enable	OFF	OFF button ends daymode
BV-66	Afterhour Timer Status	OFF	ON when AV-98 > 0, otherwise OFF
BV-67	Occupied Status	OFF	Determined by BV-64 to BV-66
BV-68	Field Service Lockout	OFF	ON = Field Service Locked out
BV-69	English/Metric display Swap	OFF	OFF = Use VLC Mode, ON = Swap
BV-70	Microset Present Flag	OFF	ON = Microset Present
BV-71	English/Metric VLC Mode	OFF	OFF = English, Read Only

Appendix C *Fire-related Input Data for Case 1_1*

File name: zfm-hvac.inp

```
ROOMS      ! Case 1_1:   Three-story building with a small fire   08/05/05
10
0.0    8.53   6.40   2.9             ! elevation, length, width, height
2.9    8.53   6.40   2.9
5.8    8.53   6.40   2.9
0.0   17.37   6.40   2.9
2.9   17.37   6.40   2.9
5.8   17.37   6.40   2.9
0.0   25.91   9.14   2.9
2.9   25.91   9.14   2.9
5.8   25.91   9.14   2.9
0.0    3.05   4.88   8.7
VENTS                                 ! number of vents
 11                                   ! from room#, to room#, bottom, top, width
  7  0 0. 2. 0.91                     ! from room 7 to outside (door)
  7  0 0. 2. 0.91
  7 10 0. 2. 0.91
  7  1 0. 2. 0.91
  7  4 0. 2. 0.91
  8 10 0. 2. 0.91
  8  2 0. 2. 0.91
  8  5 0. 2. 0.91
  9 10 0. 2. 0.91
  9  3 0. 2. 0.91
  9  6 0. 2. 0.91
SOLVEPRODS
FIRES
1                                     ! nfires
1  3 0.1 1 5.0e+7 0.002 0.5 0.2 ! room#, type, elevation, flag, heat_c,
species(i=2,3,4)
  6                                   ! number of points
     0.              0.               ! time, heat release rate(kW)
    60.            100.
    70.            200.
   100.            350.
  1800.            400.
  1900.            000.
HSIM
9                       ! number of HVAC paths
1 1.5 2.0 1 1.5 2.0     ! supply room#, duct bottom, top, return room#,
duct bottom, top
1 1 1                   ! AHU#, return path#, supply path#
2 1.5 2.0 2 1.5 2.0
1 2 2
3 1.5 2.0 3 1.5 2.0
1 3 3
4 1.5 2.0 4 1.5 2.0
2 1 1
5 1.5 2.0 5 1.5 2.0
2 2 2
6 1.5 2.0 6 1.5 2.0
2 3 3
7 1.5 2.0 7 1.5 2.0
3 1 1
8 1.5 2.0 8 1.5 2.0
```

```
3 2 2
9 1.5 2.0 9 1.5 2.0
3 3 3
PLOT
zfm-hvac
TIME
     0.0      50000.0              ! start time, end time
     0.0       2.897 300. 300.     ! pressure, layer height, lower layer temp,
upper layer temp
     0.0       2.897 300. 300.
     0.0       2.897 300. 300.
     0.0       2.897 300. 300.
     0.0       2.897 300. 300.
     0.0       2.897 300. 300.
     0.0       2.897 300. 300.
     0.0       2.897 300. 300.
     0.0       2.897 300. 300.
     0.0       8.697 300. 300.
```

Appendix D *HVAC-side Configuration Input Data*

File name: run_hvac.inp

```
    3                             ! n_ahu
    3,   3,   3                   ! nbr
AHU1.inp                          ! (1) AHU 1
AHU2.inp                          ! (2) AHU 2
AHU3.inp                          ! (3) AHU 3
VAV1.inp                          ! (4) AHU 1  (Zone 3, 2, 1)
VAV2.inp                          ! (5) AHU 2  (Zone 6, 5, 4)
VAV3.inp                          ! (6) AHU 3  (Zone 9, 8, 7)
AHU1init.inp                      ! (7) AHU 1 Initial data
AHU2init.inp                      ! (8) AHU 2 Initial data
AHU3init.inp                      ! (9) AHU 3 Initial data
AHUout1.out                       ! (10) AHU 1, 2, & 3 Output data
AHUout2.out                       ! (11) AHU 1, 2, & 3 Output data
```

Appendix E Sample AHU Input Data

File name: AHU1.inp

```
! AHU1.inp    Changed for ZFM-HVAC model
!       Base data file: emuex29i.mod

UNIT  17     TYPE 321 ----- OA damper
Motor-driven actuator

  1   INPUTS:
      CONTROL      149 - demanded position

  2   OUTPUTS:
      CONTROL       38 - position of final control element
      CONTROL        0 - motor position
      CONTROL      133 - number of stops/starts/reversals
      CONTROL        0 - total distance traveled by final control element

  3   PARAMETERS:
      1.00000    direction: 1=forward, -1=reverse, 0=stuck
      0.00000    starting position (0-1)
      125.000    travel time (lim-lim) (s)
      0.00000    minimum change in demanded position for movement (-)
      0.00000    hysteresis (-)
      0.00000    crank travel angle (0 for linear) (deg)
      1.00000    Upper limit of control element range on actuator scale
      0.00000    Lower limit of control element range on actuator scale

UNIT        TYPE ----- AHU
Mixing box (simplified)

  1   INPUTS:
      TEMPERATURE    1 - Outside air dry bulb temperature
      HUMIDITY       1 - Outside air humidity ratio
      TEMPERATURE   54 - Extract air dry bulb temperature
      HUMIDITY      16 - Extract air humidity ratio
      FLOW          29 - Extract air mass flow rate
      CONTROL       38 - Outside air damper position (0=closed, 1=open)
      OTHER          8 - incoming smoke concentration

  2   OUTPUTS:
      TEMPERATURE    2 - Mixed air dry bulb temperature
      HUMIDITY       2 - Mixed air humidity ratio
      FLOW           2 - Mixed air mass flow rate
      FLOW           1 - Total outside dry air mass flow rate
      FLOW          30 - Recirc air mass flow rate
      FLOW          47 - Exhaust air mass flow rate
      OTHER          3 - outgoing smoke concentration

  3   PARAMETERS:
      0.100000   leakage factor of zones (-)
      0.000000   Economizer on = 1, off = 0 (-)
      0.100000   Minimum damper opening (-)


UNIT  67     TYPE 321 ----- Heating coil valve
Motor-driven actuator

  1   INPUTS:
      CONTROL      159 - demanded position

  2   OUTPUTS:
      CONTROL       85 - position of final control element
      CONTROL        0 - motor position
      CONTROL       83 - number of stops/starts/reversals
      CONTROL        0 - total distance traveled by final control element
```

3 PARAMETERS:
   1.00000   direction: 1=forward, -1=reverse, 0=stuck
   0.00000   starting position (0-1)
   125.000   travel time (lim-lim) (s)
   0.00000   minimum change in demanded position for movement (-)
   0.00000   hysteresis (-)
   0.00000   crank travel angle (0 for linear) (deg)
   1.00000   Upper limit of control element range on actuator scale
   0.00000   Lower limit of control element range on actuator scale

UNIT 68   TYPE 524 ----- Heating coil and valve
Coil + three port valve, calculates flow

1 INPUTS:
   TEMPERATURE   2 - Inlet air dry bulb temperature
   HUMIDITY   2 - Inlet air humidity ratio
   PRESSURE   47 - Outlet air pressure
   FLOW   2 - Dry air mass flow rate
   TEMPERATURE   78 - Inlet water temperature
   PRESSURE   48 - Inlet water pressure
   PRESSURE   49 - Outlet water pressure
   CONTROL   85 - Valve stem position
   TEMPERATURE   79 - Effective coil surface temperature

2 OUTPUTS:
   TEMPERATURE   79 - Effective coil surface temperature
   TEMPERATURE   80 - Outlet dry bulb air temperature
   HUMIDITY   17 - Outlet air humidity ratio
   PRESSURE   0 - Inlet air pressure
   TEMPERATURE   81 - Outlet water temperature
   FLOW   0 - Primary circuit water mass flow rate
   FLOW   48 - Water mass flow rate through coil
   TEMPERATURE   0 - Mixed return water temperature
   POWER   10 - Total heat transfer to the air
   CONTROL   0 - Sensible heat ratio
   CONTROL   0 - Coil effectiveness
   CONTROL   0 - Coil by-pass factor
   ENERGY   0 - Outlet air specific enthalpy
   CONTROL   0 - Outlet air relative humidity
   TEMPERATURE   0 - Outlet air wet-bulb temperature

3 PARAMETERS:
   1.00000   Method : 0 = steady state, 1 = dynamic
   0.00000   Fault : 0 for no faults, 1 = parallel flow (cooling co
   0.00000   Psycho : 0 = no psychrometric output calcs, 1 = calcs
   1.00000   Number of rows of tubes
   18.0000   Number of tubes per row
   18.0000   Number of parallel water circuits
   0.600000E-01 Length of finned section in direction of flow (m)
   1.37000   Height of finned section (m)
   2.77000   Width of finned section (m)
   0.159000E-01 Tube outside diameter (m)
   0.787400E-03 Tube wall thickness (m)
   2.00000   Tube material (Al=1,Cu=2,Fe=3,CaCO3=4)
   0.212000E-02 Fin spacing (pitch) (m)
   0.216000E-03 Fin thickness (m)
   1.00000   Fin material (Al=1,Cu=2,Fe=3)
   0.802500E-03 Flow resistance on air side (1000/kg.m)
   0.600000   Coil water flow resistance (1000/kg.m)
   0.100000E+07 By-pass water flow resistance (1000/kg.m)
   1.00000   Valve type: 0=lin/lin, 1=eq%(flow)/lin(byp), 2=lin/eq%
   5.43600   Valve capacity (Kv) (m3/hr @ 1 bar)
   3.54000   Valve curvature parameter (0=linear) (-)
   24.5100   Valve rangability (-)
   0.100000E-03 Valve leakage (fractional flow) (-)

UNIT 20   TYPE 321 ----- Cooling coil valve
Motor-driven actuator

1 INPUTS:

             CONTROL   150 - demanded position

2   OUTPUTS:
             CONTROL   41 - position of final control element
             CONTROL   0 - motor position
             CONTROL   136 - number of stops/starts/reversals
             CONTROL   0 - total distance traveled by final control element

3   PARAMETERS:
             1.00000   direction: 1=forward, -1=reverse, 0=stuck
             0.00000   starting position (0-1)
             125.000   travel time (lim-lim) (s)
             0.00000   minimum change in demanded position for movement (-)
             0.00000   hysteresis (-)
             0.00000   crank travel angle (0 for linear) (deg)
             1.00000   Upper limit of control element range on actuator scale
             0.00000   Lower limit of control element range on actuator scale

UNIT 41   TYPE 524 ----- Cooling coil
Coil + three port valve, calculates flow

1   INPUTS:
             TEMPERATURE   80 - Inlet air dry bulb temperature
             HUMIDITY   17 - Inlet air humidity ratio
             PRESSURE   3 - Outlet air pressure
             FLOW   2 - Dry air mass flow rate
             TEMPERATURE   55 - Inlet water temperature
             PRESSURE   32 - Inlet water pressure
             PRESSURE   33 - Outlet water pressure
             CONTROL   41 - Valve stem position
             TEMPERATURE   57 - Effective coil surface temperature (same as 1st o/

2   OUTPUTS:
             TEMPERATURE   57 - Effective coil surface temperature
             TEMPERATURE   3 - Outlet dry bulb air temperature
             HUMIDITY   3 - Outlet air humidity ratio
             PRESSURE   0 - Inlet air pressure
             TEMPERATURE   56 - Outlet water temperature
             FLOW   0 - Primary circuit water mass flow rate
             FLOW   40 - Water mass flow rate through coil
             TEMPERATURE   0 - Mixed return water temperature
             POWER   3 - Total heat transfer to the air
             CONTROL   0 - Sensible heat ratio
             CONTROL   0 - Coil effectiveness
             CONTROL   0 - Coil by-pass factor
             ENERGY   0 - Outlet air specific enthalpy
             CONTROL   0 - Outlet air relative humidity
             TEMPERATURE   0 - Outlet air wet-bulb temperature

3   PARAMETERS:
             1.00000   Method : 0 = steady state, 1 = dynamic
             0.00000   Fault : 0 for no faults, 1 = parallel flow (cooling co
             0.00000   Psycho : 0 = no psychrometric output calcs, 1 = calcs
             3.00000   Number of rows of tubes
             36.0000   Number of tubes per row
             36.0000   Number of parallel water circuits
             0.180750   Length of finned section in direction of flow (m)
             1.37000   Height of finned section (m)
             2.77000   Width of finned section (m)
             0.159000E-01 Tube outside diameter (m)
             0.787400E-03 Tube wall thickness (m)
             2.00000   Tube material (Al=1,Cu=2,Fe=3,CaCO3=4)
             0.212000E-02 Fin spacing (pitch) (m)
             0.216000E-03 Fin thickness (m)
             1.00000   Fin material (Al=1,Cu=2,Fe=3)
             0.802500E-03 Flow resistance on air side (1000/kg.m)
             0.600000   Coil water flow resistance (1000/kg.m)
             0.100000E+07 By-pass water flow resistance (1000/kg.m)
             1.00000   Valve type: 0=lin/lin, 1=eq%(flow)/lin(byp), 2=lin/eq%
             16.3100   Valve capacity (Kv) (m3/hr @ 1 bar)

```
    3.54000     Valve curvature parameter (0=linear) (-)
   24.5100      Valve rangability (-)
    0.100000E-03 Valve leakage (fractional flow) (-)
```

UNIT 14   TYPE 333 ----- Supply fan
Variable speed drive

1   INPUTS:
    CONTROL   157 - demanded fractional speed

2   OUTPUTS:
    OTHER   1 - actual rotation speed

3   PARAMETERS:
      22.3330   maximum rotation speed (rev/s)
     120.000   travel time (lim-lim) (s)

UNIT 26   TYPE 350 ----- Supply fan
Fan or pump

1   INPUTS:
    FLOW   2 - mass flow rate of fluid
    PRESSURE   4 - outlet pressure
    OTHER   1 - fan or pump rotational speed

2   OUTPUTS:
    PRESSURE   3 - inlet pressure
    POWER   1 - fluid stream heat addition rate
    POWER   29 - power consumption

3   PARAMETERS:
       4.29040   1st pressure coefficient
      -1.38700   2nd pressure coefficient
       4.22930   3rd pressure coefficient
      -3.92920   4th pressure coefficient
       0.853400   5th pressure coefficient
       0.116200   1st efficiency coefficient
       1.54040   2nd efficiency coefficient
      -1.48250   3rd efficiency coefficient
       0.766400   4th efficiency coefficient
      -0.197100   5th efficiency coefficient
       0.685800   diameter (m)
       1.00000   mode: air=1, water=2
       1.00000   lowest valid normalized flow (-)
       2.00000   highest valid normalized flow (-)

UNIT 15   TYPE 333 ----- Return fan
Variable speed drive

1   INPUTS:
    CONTROL   158 - demanded fractional speed

2   OUTPUTS:
    OTHER   2 - actual rotation speed

3   PARAMETERS:
       8.00000   maximum rotation speed (rev/s)
     120.000   travel time (lim-lim) (s)

UNIT 38   TYPE 350 ----- Return fan
Fan or pump

1   INPUTS:
    FLOW   29 - mass flow rate of fluid
    PRESSURE   27 - outlet pressure
    OTHER   2 - fan or pump rotational speed

2   OUTPUTS:
    PRESSURE   26 - inlet pressure

```
        POWER        2 - fluid stream heat addition rate
        POWER       30 - power consumption

3   PARAMETERS:
        4.19370       1st pressure coefficient
       -1.63370       2nd pressure coefficient
       12.2110        3rd pressure coefficient
      -23.9619        4th pressure coefficient
        9.81620       5th pressure coefficient
        0.619000E-01  1st efficiency coefficient
        3.14170       2nd efficiency coefficient
       -5.75510       3rd efficiency coefficient
        6.16760       4th efficiency coefficient
       -3.37480       5th efficiency coefficient
        1.13000       diameter (m)
        1.00000       mode: air=1, water=2
        0.700000      lowest valid normalized flow (-)
        1.00000       highest valid normalized flow (-)
```

UNIT 69   TYPE 341 ----- AHU HC
Fluid resistance

1   INPUTS:
        FLOW   2 - Mass flow rate
        PRESSURE   47 - Outlet pressure

2   OUTPUTS:
        PRESSURE   2 - Inlet pressure

3   PARAMETERS:
        0.200000E-03 flow resistance (0.001/kg.m)

UNIT 25   TYPE 341 ----- AHU duct and CC
Fluid resistance

1   INPUTS:
        FLOW   2 - Mass flow rate
        PRESSURE   3 - Outlet pressure

2   OUTPUTS:
        PRESSURE   47 - Inlet pressure

3   PARAMETERS:
        0.546000E-03 flow resistance (1000/kg.m)

UNIT 27   TYPE 341 ----- Supply duct
Fluid resistance

1   INPUTS:
        FLOW   2 - Mass flow rate
        PRESSURE   5 - Outlet pressure

2   OUTPUTS:
        PRESSURE   4 - Inlet pressure

3   PARAMETERS:
        0.164000E-03 flow resistance (1000/kg.m)

UNIT 39   TYPE 341 ----- Return duct
Fluid resistance

1   INPUTS:
        FLOW   29 - Mass flow rate
        PRESSURE   28 - Outlet pressure

2   OUTPUTS:
        PRESSURE   27 - Inlet pressure

3   PARAMETERS:
    0.118000E-03   flow resistance (1000/kg.m)

UNIT 42   TYPE 366 ----- AHU
Ideal heating of fluid stream

1   INPUTS:
    FLOW   2 - mass flow rate
    TEMPERATURE   3 - inlet temperature
    POWER   1 - rate of heat addition

2   OUTPUTS:
    TEMPERATURE   4 - outlet temperature

3   PARAMETERS:
    1.00000   fluid: 1 = air, any other value + water (-)
    30.0000   time constant (s)

UNIT 47   TYPE 366 ----- AHU return stream
Ideal heating of fluid stream

1   INPUTS:
    FLOW   29 - mass flow rate
    TEMPERATURE   53 - inlet temperature
    POWER   2 - rate of heat addition

2   OUTPUTS:
    TEMPERATURE   54 - outlet temperature

3   PARAMETERS:
    1.00000   fluid: 1 = air, any other value + water (-)
    30.0000   time constant (s)

UNIT 62   TYPE 305 ----- Supply air
Static pressure sensor

1   INPUTS:
    PRESSURE   5 - pressure
    FLOW   2 - mass flow rate

2   OUTPUTS:
    CONTROL   65 - sensor output

3   PARAMETERS:
    1.00000   air = 1, water = 2 (-)
    1.00000   cross-sectional area of duct or pipe (m2)
    0.00000   offset: input for zero output (kPA)
    1.00000   gain: change in input for 0->1 output (kPA)
    1.00000   time constant (s)
    5.00000   upper limit of output range (-)
    -5.00000   lower limit of output range (-)

UNIT 60   TYPE 303 ----- Supply air
Flow rate sensor

1   INPUTS:
    FLOW   2 - mass flow rate

2   OUTPUTS:
    CONTROL   63 - sensor output

3   PARAMETERS:
    2.00000   mass flow=1, vol flow=2, vel=3, vel pres=4 (-)
    1.00000   air = 1, water = 2 (-)
    1.00000   cross-sectional area of duct or pipe (m2)
    0.00000   offset: input for zero output (sensed quantity)

```
        1.00000   gain: change in input for 0->1 output (sensed quantity)
        1.00000   time constant (s)
       20.0000    upper limit of output range (-)
      -10.0000    lower limit of output range (-)
```

UNIT 61   TYPE 303 ----- Return air
Flow rate sensor

1   INPUTS:
       FLOW   29 - mass flow rate

2   OUTPUTS:
       CONTROL   64 - sensor output

3   PARAMETERS:
```
        2.00000   mass flow=1, vol flow=2, vel=3, vel pres=4 (-)
        1.00000   air = 1, water = 2 (-)
        1.00000   cross-sectional area of duct or pipe (m2)
        0.00000   offset: input for zero output (sensed quantity)
        1.00000   gain: change in input for 0->1 output (sensed quantity)
        1.00000   time constant (s)
       20.0000    upper limit of output range (-)
      -10.0000    lower limit of output range (-)
```

UNIT 54   TYPE 301 ----- SA
Temperature sensor

1   INPUTS:
       TEMPERATURE   4 - temperature

2   OUTPUTS:
       CONTROL   61 - sensor output

3   PARAMETERS:
```
        0.00000    offset: input for zero output (C)
        1.00000    gain: change in input for 0->1 output (K)
       30.0000     time constant (s)
     2000.0000     upper limit of output range (-)
        0.00000    lower limit of output range (-)
```

UNIT 56   TYPE 301 ----- MA
Temperature sensor

1   INPUTS:
       TEMPERATURE   2 - temperature

2   OUTPUTS:
       CONTROL   68 - sensor output

3   PARAMETERS:
```
        0.00000    offset: input for zero output (C)
        1.00000    gain: change in input for 0->1 output (K)
       30.0000     time constant (s)
     2000.0000     upper limit of output range (-)
        0.00000    lower limit of output range (-)
```

UNIT 53   TYPE 301 ----- OA
Temperature sensor

1   INPUTS:
       TEMPERATURE   1 - temperature

2   OUTPUTS:
       CONTROL   60 - sensor output

3   PARAMETERS:
```
        0.00000    offset: input for zero output (C)
        1.00000    gain: change in input for 0->1 output (K)
```

            30.0000   time constant (s)
            51.6700   upper limit of output range (-)
           -40.0000   lower limit of output range (-)

UNIT 55   TYPE 301 ----- RA
Temperature sensor

1   INPUTS:
         TEMPERATURE   54 - temperature

2   OUTPUTS:
         CONTROL   62 - sensor output

3   PARAMETERS:
         0.00000   offset: input for zero output (C)
         1.00000   gain: change in input for 0->1 output (K)
        30.0000   time constant (s)
      2000.0000   upper limit of output range (-)
         0.00000   lower limit of output range (-)

UNIT 63   TYPE 302 ----- Outside air
Humidity sensor

1   INPUTS:
         HUMIDITY   1 - humidity ratio
         TEMPERATURE   1 - temperature

2   OUTPUTS:
         CONTROL   66 - sensor output

3   PARAMETERS:
         1.00000   HR=1, RH=2, DP=3, WB=4, deg sat=5, spec enthalpy=6 (-)
         0.00000   offset: input for zero output (sensed quantity)
         1.00000   gain: change in input for 0->1 output (sensed quantity)
        30.0000   time constant (s)
         1.00000   upper limit of output range (-)
         0.00000   lower limit of output range (-)

UNIT 64   TYPE 302 ----- Return air
Humidity sensor

1   INPUTS:
         HUMIDITY   16 - humidity ratio
         TEMPERATURE   54 - temperature

2   OUTPUTS:
         CONTROL   67 - sensor output

3   PARAMETERS:
         1.00000   HR=1, RH=2, DP=3, WB=4, deg sat=5, spec enthalpy=6 (-)
         0.00000   offset: input for zero output (sensed quantity)
         1.00000   gain: change in input for 0->1 output (sensed quantity)
        30.0000   time constant (s)
         1.00000   upper limit of output range (-)
         0.00000   lower limit of output range (-)

Appendix E *Sample VAV Input Data*

File name: VAV1.inp

```
! VAV1.inp  Changed for ZFM-HVAC model
! Base data file: emuex29i.mod

UNIT 30     TYPE 526 ----- zone 1
Motorized pressure-independent VAV box

1   INPUTS:
    PRESSURE     5 - inlet pressure
    PRESSURE    16 - outlet pressure
    CONTROL    151 - demanded flow rate (normalized to nominal)
    FLOW         3 - air mass flow rate

2   OUTPUTS:
    FLOW         3 - air mass flow rate
    CONTROL     77 - damper position (0=closed, 1=open)
    CONTROL      0 - fractional motor velocity
    CONTROL    143 - number of stops/starts/reversals
    CONTROL      0 - total distance traveled by valve/damper

3   PARAMETERS:
    0.565000      nominal volumetric flow rate (m3/s)
    0.250000      face area of damper(s) (m2)
    0.999000E-01  additional pressure drop at nominal flow (kPa)
    100.000       actuator travel time (0-90 deg) (s)
    0.00000       minimum fractional motor speed (-)
    0.00000       hysteresis (-)
    9.99000       controller gain (frac speed/frac error)

UNIT 21     TYPE 321 ----- RHC valve zone 1
Motor-driven actuator

1   INPUTS:
    CONTROL    154 - demanded position

2   OUTPUTS:
    CONTROL     42 - position of final control element
    CONTROL      0 - motor position
    CONTROL    137 - number of stops/starts/reversals
    CONTROL      0 - total distance traveled by final control element

3   PARAMETERS:
    1.00000    direction: 1=forward, -1=reverse, 0=stuck
    0.00000    starting position (0-1)
    125.000    travel time (lim-lim) (s)
    0.00000    minimum change in demanded position for movement (-)
    0.00000    hysteresis (-)
    0.00000    crank travel angle (0 for linear) (deg)
    1.00000    Upper limit of control element range on actuator scale
    0.00000    Lower limit of control element range on actuator scale

UNIT 43     TYPE 524 ----- zone 1
Coil + three port valve, calculates flow

1   INPUTS:
    TEMPERATURE    4 - Inlet air dry bulb temperature
    HUMIDITY       3 - Inlet air humidity ratio
    PRESSURE      16 - Outlet air pressure
    FLOW           3 - Dry air mass flow rate
    TEMPERATURE   58 - Inlet water temperature
    PRESSURE      34 - Inlet water pressure
    PRESSURE      35 - Outlet water pressure
    CONTROL       42 - Valve stem position
    TEMPERATURE   70 - Effective coil surface temperature (same as 1st o/
```

2 OUTPUTS:
   TEMPERATURE   70 - Effective coil surface temperature
   TEMPERATURE   5 - Outlet dry bulb air temperature
   HUMIDITY   4 - Outlet air humidity ratio
   PRESSURE   0 - Inlet air pressure
   TEMPERATURE   64 - Outlet water temperature
   FLOW   0 - Primary circuit water mass flow rate
   FLOW   41 - Water mass flow rate through coil
   TEMPERATURE   0 - Mixed return water temperature
   POWER   4 - Total heat transfer to the air
   CONTROL   0 - Sensible heat ratio
   CONTROL   0 - Coil effectiveness
   CONTROL   0 - Coil by-pass factor
   ENERGY   0 - Outlet air specific enthalpy
   CONTROL   0 - Outlet air relative humidity
   TEMPERATURE   0 - Outlet air wet-bulb temperature

3 PARAMETERS:
   1.00000   Method : 0 = steady state, 1 = dynamic
   0.00000   Fault : 0 for no faults, 1 = parallel flow (cooling co
   0.00000   Psycho : 0 = no psychrometric output calcs, 1 = calcs
   1.00000   Number of rows of tubes
   16.0000   Number of tubes per row
   1.00000   Number of parallel water circuits
   0.275000E-01   Length of finned section in direction of flow (m)
   0.457000   Height of finned section (m)
   0.305000   Width of finned section (m)
   0.127000E-01   Tube outside diameter (m)
   0.406000E-03   Tube wall thickness (m)
   2.00000   Tube material (Al=1,Cu=2,Fe=3,CaCO3=4)
   0.254000E-02   Fin spacing (pitch) (m)
   0.152000E-03   Fin thickness (m)
   1.00000   Fin material (Al=1,Cu=2,Fe=3)
   0.150000   Flow resistance on air side (0.001 kg.m)
   25.0000   Coil water flow resistance (0.001 kg.m)
   0.100000E+07   By-pass water flow resistance (0.001 kg.m)
   1.00000   Valve type: 0=lin/lin, 1=eq%(flow)/lin(byp), 2=lin/eq%
   0.500000   Valve capacity (Kv) (m3/hr @ 1 bar)
   3.22000   Valve curvature parameter (0=linear) (-)
   50.0000   Valve rangability (-)
   0.200000E-03   Valve leakage (fractional flow) (-)

UNIT 33   TYPE 425 ----- zone 1
Room air mass balance

1 INPUTS:
   FLOW   3 - supply air mass flow rate (positive in)
   PRESSURE   16 - room pressure
   PRESSURE   30 - ambient pressure
   TEMPERATURE   35 - room temperature

2 OUTPUTS:
   FLOW   19 - return air mass flow rate (positive out)
   FLOW   31 - leakage mass flow rate to ambient (positive out)

3 PARAMETERS:
   1.00000   leakage resistance [0.001/(kg m)]
   0.00000   local extract fan mass flow rate [kg/s]

UNIT 57   TYPE 301 ----- zone 1 air temperature
Temperature sensor

1 INPUTS:
   TEMPERATURE   35 - temperature

2 OUTPUTS:
   CONTROL   70 - sensor output

3 PARAMETERS:
   0.00000   offset: input for zero output (C)

    1.00000   gain: change in input for 0->1 output (K)
    10.0000   time constant (s)
    2000.00   upper limit of output range (-)
    0.00000   lower limit of output range (-)

UNIT 37   TYPE 422 ----- From Zone 1
Flow merge with smoke

3   PARAMETERS:
    0.263200E-03 Inlet resistance 1 (0.001/k.m)
    0.263200E-03 Inlet resistance 2 (0.001/k.m)
    0.263200E-02 Resistance of outlet (0.001/k.m)

UNIT 46   TYPE 365 ----- AHU
Mixing of two moist air streams

3   PARAMETERS:
    0.00000   (dummy)

Appendix G Sample AHU Initial Data

File name: AHU1init.inp

```
PRESSURE         1 ->      0.00000      (kPa)
PRESSURE         2 ->     -0.09278      (kPa)
PRESSURE         3 ->     -0.09375      (kPa)
PRESSURE         4 ->      0.25100      (kPa)
PRESSURE         5 ->      0.25000      (kPa)
PRESSURE        16 ->      0.01000      (kPa)
PRESSURE        26 ->     -0.46000      (kPa)
PRESSURE        27 ->      0.06465      (kPa)
PRESSURE        28 ->      0.06465      (kPa)
PRESSURE        29 ->      0.06465      (kPa)
PRESSURE        30 ->      0.00000      (kPa)
PRESSURE        32 ->     39.0300       (kPa)
PRESSURE        33 ->      0.00000      (kPa)
PRESSURE        34 ->     38.1000       (kPa)
PRESSURE        35 ->      0.00000      (kPa)
PRESSURE        47 ->      0.00000      (kPa)
PRESSURE        48 ->     20.0000       (kPa)
PRESSURE        49 ->      0.00000      (kPa)
FLOW             1 ->      0.66100      (kg/s)
FLOW             2 ->      1.33000      (kg/s)
FLOW             3 ->      0.28300      (kg/s)
FLOW            19 ->      0.31200      (kg/s)
FLOW            28 ->      0.00000      (kg/s)
FLOW            29 ->      1.10000      (kg/s)
FLOW            30 ->      0.67300      (kg/s)
FLOW            31 ->      0.00000      (kg/s)
FLOW            40 ->      0.00000      (kg/s)
FLOW            41 ->      0.00000      (kg/s)
FLOW            47 ->      0.00000      (kg/s)
FLOW            48 ->      0.00000      (kg/s)
TEMPERATURE      1 ->     30.0000       (C)
TEMPERATURE      2 ->     20.0000       (C)
TEMPERATURE      3 ->     20.0000       (C)
TEMPERATURE      4 ->     20.0000       (C)
TEMPERATURE      5 ->     20.0000       (C)
TEMPERATURE     35 ->     25.0000       (C)
TEMPERATURE     53 ->     20.0000       (C)
TEMPERATURE     54 ->     20.0000       (C)
TEMPERATURE     55 ->      6.00000      (C)
TEMPERATURE     56 ->      6.00000      (C)
TEMPERATURE     57 ->     20.0000       (C)
TEMPERATURE     58 ->     80.0000       (C)
TEMPERATURE     64 ->     20.0000       (C)
TEMPERATURE     70 ->     20.0000       (C)
TEMPERATURE     78 ->     80.0000       (C)
TEMPERATURE     79 ->     20.0000       (C)
TEMPERATURE     80 ->     20.0000       (C)
TEMPERATURE     81 ->     20.0000       (C)
CONTROL          1 ->      1.0000       (-)
CONTROL          2 ->      1.0000       (-)
CONTROL         38 ->      0.0000       (-)
CONTROL         41 ->      0.0000       (-)
CONTROL         42 ->      0.0000       (-)
CONTROL         60 ->     30.0000       (-)
CONTROL         61 ->     20.0000       (-)
CONTROL         62 ->     20.0000       (-)
CONTROL         63 ->      6.17300      (-)
CONTROL         64 ->      5.25600      (-)
CONTROL         65 ->      0.223000     (-)
CONTROL         66 ->      0.153000E-01 (-)
CONTROL         67 ->      0.750000E-02 (-)
CONTROL         68 ->     20.0000       (-)
CONTROL         70 ->     22.2200       (-)
CONTROL         77 ->      0.00000      (-)
```

```
CONTROL         85 ->    0.00000       (-)
CONTROL        143 ->    0.00000       (-)
CONTROL        149 ->    0.10000       (-)
CONTROL        150 ->    0.26000       (-)
CONTROL        151 ->    0.70000       (-)
CONTROL        154 ->    0.00000       (-)
CONTROL        157 ->    0.50000       (-)
CONTROL        158 ->    0.20000       (-)
CONTROL        159 ->    0.00000       (-)
OTHER            3 ->    0.00000       (-)
OTHER            4 ->    0.00000       (-)
OTHER            8 ->    0.00000       (-)
POWER            1 ->    0.00000       (kW)
POWER            2 ->    0.00000       (kW)
POWER            3 ->    0.00000       (kW)
POWER            4 ->    0.00000       (kW)
POWER           10 ->    0.00000       (kW)
POWER           29 ->    0.00000       (kW)
POWER           30 ->    0.00000       (kW)
HUMIDITY         1 ->    0.162000E-01  (kg/kg)
HUMIDITY         2 ->    0.162000E-01  (kg/kg)
HUMIDITY         3 ->    0.162000E-01  (kg/kg)
HUMIDITY         4 ->    0.162000E-01  (kg/kg)
HUMIDITY        10 ->    0.162000E-01  (kg/kg)
HUMIDITY        16 ->    0.162000E-01  (kg/kg)
HUMIDITY        17 ->    0.162000E-01  (kg/kg)
```

www.ingramcontent.com/pod-product-compliance
Lightning Source LLC
Chambersburg PA
CBHW081730170526
45167CB00009B/3768